SINC METHODS
for Quadrature and Differential Equations

SINC METHODS
for Quadrature and Differential Equations

John Lund
Kenneth L. Bowers
Montana State University

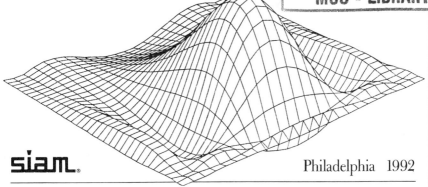

siam.

Philadelphia 1992

Society for Industrial and Applied Mathematics

Library of Congress Cataloging-in-Publication Data

Lund, J. (John)
　　Sinc methods for quadrature and differential equations/John Lund
and Kenneth L. Bowers.
　　　p.　cm.
　　Includes bibliographical references and index.
　　ISBN 0-89871-298-X
　　1. Galerkin methods. 2. Numerical integration. 3. Differential
equations—Numerical solutions.　　I. Bowers, K. (Kenneth)
II. Title.
QA372.L86 1992
515'.35—dc20   92-12139

All rights reserved. Printed in the United States of America. No part of this book may be reproduced, stored, or transmitted in any manner without the written permission of the Publisher. For information, write the Society for Industrial and Applied Mathematics, 3600 University City Science Center, Philadelphia, Pennsylvania 19104-2688.

Copyright ©1992 by the Society for Industrial and Applied Mathematics.

siam. is a registered trademark.

*To my parents,
John and Esther,
for many reasons.*
— *John Lund*

*To my wife,
Sandy,
whose love has sustained me
and to my children,
Kevin and Cheryl,
whose joyful hearts have
lightened my own.*
— *Ken Bowers*

Contents

Preface		ix
Chapter 1.	*Preliminary Material*	**1**
	1.1 Analytic Functions	2
	1.2 Conformal Mapping	7
	1.3 Fourier Series and Transforms	9
Chapter 2.	*Numerical Methods on the Real Line*	**21**
	2.1 Exact Interpolation and Quadrature	22
	2.2 Approximate Interpolation and Quadrature	33
Chapter 3.	*Numerical Methods on an Arc* Γ	**57**
	3.1 Interpolation on Γ	58
	3.2 Quadrature on Γ	68
	3.3 Differentiation on Γ	92
Chapter 4.	*The Sinc-Galerkin Method*	**101**
	4.1 A Galerkin Scheme	102
	4.2 Inner Product Approximations	104
	4.3 Discrete System Assembly and Error	113
	4.4 Treatment of Boundary Conditions	140
	4.5 Analytic Considerations	151
Chapter 5.	*Steady Problems*	**163**
	5.1 Sinc-Collocation Methods	164
	5.2 Sturm–Liouville Problems	174
	5.3 Discretization of a Self-Adjoint Form	188
	5.4 The Poisson Problem	194
Chapter 6.	*Time-Dependent Problems*	**227**
	6.1 The Heat Equation	228
	6.2 Treatment of Boundary and Initial Conditions	242
	6.3 Burgers' Equation	257

Appendix A. *Linear Algebra* — 269
    A.1 Special Matrices and Diagonalization — 270
    A.2 Eigenvalue Bounds — 277
    A.3 Block Matrices — 281

Bibliography — 291

Text Directory — 299

Index — 303

Preface

This book focuses on the exploration of the route that leads to the Sinc-Galerkin method for solving time-dependent partial differential equations. Although the aim of the book is ultimately singular in focus, the presentation of the material allows the reader to wander down a few peripheral paths.

Along the way to the final chapter, in which the fully Sinc-Galerkin method is thoroughly addressed, one encounters a powerful method for solving ordinary differential equations. This power is especially apparent in the method's adaptability to problems with singular solutions. Indeed, in a list of the sinc function's many elegant approximate properties, its defiant disregard of boundary singularities probably outranks them all.

Because Galerkin methods require quadratures, the text covers all the sinc integration rules from an analytic, as well as a practical, viewpoint. The quadrature rules are conveniently obtained via interpolation. Hence the interpolation formulas are developed for the purpose of the quadrature method, as well as for their use in the sinc-collocation procedures. The interpolation formulas are obtained from Cauchy's theorem and conformal mappings abound throughout the book, so the opening chapter includes review material from analytic function theory.

Sinc function methodology is a relatively new entry in the toolbox of numerical techniques, so the time seemed right to present an accessible, more elementary version of the material than is currently found in the literature. Using differential equations as a backdrop for the material afforded us the opportunity to present a number of the applications of the sinc method that have proved useful in numerical processes. These applications are of interest quite independent of differential equations.

The contents of this book have been taught to graduate classes at both Montana State University, Bozeman, Montana and Texas Tech University, Lubbock, Texas. The original idea was to expose the students to a class of numerical methods which complement the techniques they learn in a standard course in numerical analysis. These students were patient and perceptive with regard to the presented material, the examples, and the problems. Only time will tell if we have struck an appropriate balance. The long-term health and development of any discipline depend on the students of the discipline. To our delight, some of these students chose to pursue the study of sinc functions. It is to these students that we express our fondest appreciation.

A book is rarely the sole product of the authors. This text is no exception. The patient advice and editorial assistance we received from the ladies of SIAM was more valuable than Swedish Crystal. The other help we received in preparing this manuscript ranged from the skilled and the technical to the inspirational. An unfulfilled accounting goes to: René Tritz, whose fingers dance a Diamond note on the keyboard; to Tim Carlson, who could fish a PostScript file out of a stream of disconnections; and to Professor Frank Stenger, who took E.T. Whittaker's function of "royal blood," calculated with it, and crowned it a prince.

Chapter 1

Preliminary Material

A convenient beginning for studying the method of sinc approximation is obtained via the residue theorem of complex analysis. This theorem and a few other results from analytic function theory are collected in Section 1.1. Conformally equivalent domains play a fundamental role throughout the text, so this topic is briefly reviewed in Section 1.2. The final Section 1.3 sets the notation and definition for Fourier series and transforms. The latter leads to a statement of the very important Paley–Wiener theorem. The ideas are illustrated with examples that expose various analytic properties of the sinc function. All of the material in this chapter can be found in many standard references such as [1], [3], [4], [5], and [6].

1.1 Analytic Functions

Introductory material from complex analysis is collected in the following definitions and theorems. Standard notation is introduced, and the basic concepts of a domain and a simple closed contour are defined. Finally, the fundamental contour integral theorems are stated.

Let $\mathbb{R}$ denote the set of *real numbers*, and for the set of *complex numbers* write

$$\mathbb{C} \equiv \{x + iy : x, y \in \mathbb{R} \text{ and } i = \sqrt{-1}\,\}.$$

The extended complex plane is denoted $\overline{\mathbb{C}} \equiv \mathbb{C} \cup \{\infty\}$. If $z = x + iy \in \mathbb{C}$, then the *real* and *imaginary parts* of z are denoted by $Re(z) \equiv x$ and $Im(z) \equiv y$, respectively. The *complex conjugate* of z is $\bar{z} \equiv x - iy$. The *modulus* and *argument* of z are given in terms of the polar form of the complex number

$$z = r \exp(i\theta) = r(\cos(\theta) + i\sin(\theta))$$

where $r = |z| \equiv (z\bar{z})^{1/2} = (x^2 + y^2)^{1/2}$ and $\arg(z) \equiv \theta$. The angle θ is determined up to a multiple of 2π and throughout the text will be restricted to $-\pi < \theta \leq \pi$. With this restriction, and assuming that $z \neq 0$, the angle θ is given in terms of z by $\theta = \tan^{-1}(y/x)$, $z \neq 0$.

Definition 1.1 An open set $S \subseteq \mathbb{C}$ is called *connected* if it cannot be written as the union of two disjoint open sets A and B such that both A and B intersect S. An open set $S \subseteq \mathbb{C}$ is called *simply connected* if $\overline{\mathbb{C}} \setminus S$ is connected. A *domain* D is an open simply connected subset of $\mathbb{C}$.

Definition 1.2 A *smooth arc* is a set of points in $\mathbb{C}$, $z(t) = x(t) + iy(t)$, $a \leq t \leq b$, such that $z \in C^1[a, b]$ and $z'(t) \neq 0$ for any $t \in [a, b]$. The arc is *piecewise differentiable* if it is differentiable except at a finite set of points where it is continuous. The arc is called a *closed path* if $z(a) = z(b)$. A *simple closed contour* γ in $\mathbb{C}$ given by $z(t) = x(t) + iy(t)$, $a \leq t \leq b$, is a piecewise differentiable closed path with the only intersection $z(a) = z(b)$.

Definition 1.3 A function f is called *differentiable* at $z_0 \in \mathbb{C}$ if

$$f'(z_0) \equiv \lim_{z \to z_0} \left[\frac{f(z) - f(z_0)}{z - z_0} \right]$$

exists. $f'(z_0)$ is called the *derivative* of f at z_0. A function f is *analytic* at z_0 if its derivative exists at each point in some neighborhood of z_0. The function f is *analytic in a domain* D if it is analytic at each point in D. If f is analytic in all of $\mathbb{C}$, f is called *entire*.

Theorem 1.4 (Cauchy's Theorem) *If γ is a simple closed contour in a domain D and f is analytic in D, then*

$$\int_\gamma f(z) dz = 0.$$

∎

Theorem 1.5 (Cauchy's Integral Formula) *Assume that γ is a simple closed contour in D, z is contained in the interior of γ, and f is analytic in D. Then*

$$f(z) = \frac{1}{2\pi i} \int_\gamma \frac{f(w) dw}{(w - z)}$$

where γ is positively oriented. Positively oriented means

$$\frac{1}{2\pi i} \int_\gamma \frac{dw}{(w - z)} = 1.$$

Geometrically, γ is traversed in the counterclockwise direction.

∎

Theorem 1.6 (Morera's Theorem) *If f is a continuous function in D and for every simple closed contour γ in D,*

$$\int_\gamma f(z) dz = 0,$$

then f is analytic in D.

∎

Theorem 1.7 (Laurent's Theorem) *If f is analytic in the annular region $A = \{z \in \mathbb{C} : r < |z - z_0| < R\}$, where $0 \leq r < R \leq \infty$, then the function f has the Laurent series representation*

$$f(z) = \sum_{n=-\infty}^{\infty} a_n (z - z_0)^n \qquad (1.1)$$

where

$$a_n = \frac{1}{2\pi i} \int_\gamma \frac{f(w)\,dw}{(w - z_0)^{n+1}}. \qquad (1.2)$$

Here γ is a positively oriented simple closed contour in A containing z_0. The Laurent series converges absolutely on A and uniformly on any set $B = \{z \in \mathbb{C} : r < \rho_1 \leq |z - z_0| \leq \rho_2 < R\}$.

∎

Definition 1.8 If f is analytic in the region $R = \{z \in \mathbb{C} : 0 < |z - z_0| < r\}$ but not analytic at z_0, then f has an *isolated singularity* at z_0. This singularity is classified in terms of the numbers a_{-n}, for $n > 0$ in (1.2). The singularity is termed:

(a) *removable* if $a_{-n} = 0$ for all $n = 1, 2, \ldots$.

(b) a *pole of order* k if, for some $k \geq 1$, $a_{-k} \neq 0$, and for every $n > k$, $a_{-n} = 0$.

(c) *essential* if $a_{-n} \neq 0$ for infinitely many indices $n \geq 1$.

The number a_{-1} is called the *residue* of f at z_0. The residue will be denoted $\mathrm{Res}(f, z_0)$, so with (1.2)

$$\mathrm{Res}(f, z_0) \equiv a_{-1} = \frac{1}{2\pi i} \int_\gamma f(w)\,dw$$

where γ is a positively oriented simple closed contour in $A = \{z \in \mathbb{C} : r < |z - z_0| < R\}$ containing z_0.

Example 1.9 The functions $\sin(\pi z)/(\pi z)$, $\exp(z)/z$, and $\exp(1/z)$ have, respectively, a removable singularity, a pole of order one, and an essential singularity at $z_0 = 0$. Furthermore, these functions have residues 0, 1, and 1, respectively, at $z_0 = 0$. In the case of a removable

PRELIMINARY MATERIAL

singularity (as in the first function), if the function is defined to be its limiting value at the removable singularity, the extended function becomes analytic there. Example 1.10 provides a direct verification that $\sin(\pi z)/(\pi z)$ has a removable singularity at $z_0 = 0$.

∎

Example 1.10 Let $f(z) = \sin(\pi z)/(\pi z)$ and $z_0 = 0$. Then $a_{-n} = 0$ for all $n = 1, 2, \ldots$ in Laurent's Theorem 1.7. To see this using (1.2), let $\gamma = \{w \in \mathbb{C} : |w| = \rho\}$ and write

$$a_{-n} = \frac{1}{2\pi i} \int_{|w|=\rho} \frac{\sin(\pi w)}{\pi w^{-n+2}} dw.$$

For $n \geq 2$ the integrand is analytic and the result follows from Cauchy's Theorem 1.4. For $n = 1$, parameterize the curve $|w| = \rho$ by $w = \rho(\cos\theta + i\sin\theta)$, $-\pi < \theta \leq \pi$. Then

$$a_{-1} = \frac{1}{2\pi^2} \int_{-\pi}^{\pi} \sin[\pi\rho(\cos\theta + i\sin\theta)] d\theta.$$

Expanding and noting that $\sinh(x)$ is an odd function yields

$$\begin{aligned} a_{-1} &= \frac{1}{\pi^2} \int_0^{\pi} \sin[\rho\pi \cos\theta] \cosh[\rho\pi \sin\theta] d\theta \\ &= \frac{1}{\pi^2} \int_0^{\pi/2} \sin[\rho\pi \cos\theta] \cosh[\rho\pi \sin\theta] d\theta \\ &+ \frac{1}{\pi^2} \int_{\pi/2}^{\pi} \sin[\rho\pi \cos\theta] \cosh[\rho\pi \sin\theta] d\theta. \end{aligned}$$

The substitution $\phi = \pi - \theta$ in the second integral shows that $a_{-1} = \text{Res}(f, 0) = 0$. Hence, $a_{-n} = 0$ for all $n = 1, 2, \ldots$ and the singularity is removable. Upon defining $f(0) = 1$, the function f is analytic at $z_0 = 0$.

∎

Definition 1.11 The function defined for all $z \in \mathbb{C}$ by

$$\text{sinc}(z) \equiv \begin{cases} \dfrac{\sin(\pi z)}{\pi z}, & z \neq 0 \\ 1, & z = 0 \end{cases} \tag{1.3}$$

is called the *sinc function*. By Example 1.10, this is an entire function.

The computation that took place in Example 1.10 indicates that a method different from the definition of a residue via the contour integral would be desirable for such calculations.

Theorem 1.12 *Let D be a domain. If f is analytic in $D \setminus \{z_0\}$ and has a pole of order $k+1$ at z_0, then*

$$Res(f, z_0) = \frac{1}{k!} \lim_{z \to z_0} \left[\frac{d^k}{dz^k} \left[(z-z_0)^{k+1} f(z) \right] \right]. \tag{1.4}$$

If $f = g/h$, where g and h are analytic at z_0 with $g(z_0) \neq 0$, $h(z_0) = 0$, and $h'(z_0) \neq 0$, then f has a pole of order one (a simple pole) and

$$Res(f, z_0) = \frac{g(z_0)}{h'(z_0)}. \tag{1.5}$$

∎

Theorem 1.13 (Residue Theorem) *Let D be a domain. If f is analytic in $D \setminus \{z_j\}_{j=1}^n$ and $\{z_j\}_{j=1}^n$ is contained in the interior of the simple closed contour $\gamma \subset D$, then*

$$\int_\gamma f(z) dz = 2\pi i \sum_{j=1}^n Res(f, z_j).$$

∎

Example 1.14 Let h and d be positive constants. Assume f is analytic in a domain D containing $B_k \cup \partial B_k$, where for each positive integer k the boundary of the box B_k is denoted ∂B_k and

$$B_k = \{z = x + iy : (k - (1/2))h < x < (k + (1/2))h, \, |y| < d\}.$$

The function

$$g(z) = \frac{\sin(\pi x/h) f(z)}{(z-x) \sin(\pi z/h)}$$

satisfies the following equations for $x \neq kh$:

$$Res(g, kh) = \frac{(-1)^k h \sin(\pi x/h) f(kh)}{\pi (kh - x)} \tag{1.6}$$

and
$$Res(g,x) = f(x). \tag{1.7}$$

These follow from (1.5). Furthermore, if $x \neq kh$, then the Residue Theorem 1.13 yields, from the singularities at $z = x$ and $z = kh$,

$$\begin{aligned}\int_{\partial B_k} g(z)dz &= 2\pi i \left[Res(g,x) + Res(g,kh)\right] \\ &= 2\pi i \left[f(x) + \frac{(-1)^k h \sin(\pi x/h) f(kh)}{\pi(kh-x)}\right].\end{aligned}$$

Minor rearrangement then gives the identity

$$f(x) = f(kh)\operatorname{sinc}\left(\frac{x-kh}{h}\right) + \frac{\sin(\pi x/h)}{2\pi i}\int_{\partial B_k} \frac{f(z)dz}{(z-x)\sin(\pi z/h)}.$$

This equation represents a one-point interpolation formula (with error) for the function f at the point $x = kh$.

∎

1.2 Conformal Mapping

Many problems that arise in applied mathematics do not have the whole real line as their natural domain. Given a problem on an interval and a numerical process on the real line then, roughly speaking, there are two points of view. One is to change variables in the problem so that, in the new variables, the problem has a domain corresponding to that of the numerical process. A second procedure is to move the numerical process and to study it on the new domain. The latter approach is the method chosen here. The development for transforming sinc methods from one domain to another is accomplished via conformal mappings.

Definition 1.15 If f is analytic in a domain D and $z_0 \in D$, then f is called *conformal* at z_0 if $f'(z_0) \neq 0$. If $f'(z) \neq 0$ for all $z \in D$, then f is called a *conformal mapping of* D.

The geometric implications of the previous definition are that if the mapping f is conformal at z_0, then the angle between two arcs through z_0 is the same as the angle between their images under f at $f(z_0)$; the orientation of the angle is thus preserved.

Example 1.16 The mapping

$$\phi(z) = \ell n(z) = \ell n|z| + i\arg(z), \quad |\arg(z)| < d \leq \pi/2$$

is a conformal map of the domain

$$D_W = \{z \in \mathbb{C} : z = r\exp(i\theta),\ r > 0,\ |\theta| < d\}$$

onto the domain (Figure 1.1)

$$D_S = \{w \in \mathbb{C} : w = t + iv,\ t \in \mathbb{R},\ |v| < d\}$$

since $\phi(z)$ is analytic on D_W and $\phi'(z) = \frac{1}{z} \neq 0$ for all $z \in D_W$.

Figure 1.1 The conformal map $\phi(z) = \ell n(z)$ of the domain D_W onto the domain D_S.

■

A special class of conformal maps are given by the linear fractional transformation

$$L(z) = \frac{az+b}{cz+d}, \quad ad - bc \neq 0 \tag{1.8}$$

which is a conformal mapping of $\overline{\mathbb{C}}$ onto $\overline{\mathbb{C}}$. Geometrically, linear fractional transformations preserve angles. Furthermore if straight lines are regarded as circles with an infinite radius, then they map circles to circles. This property is helpful in Exercise 1.1.

Several properties of conformal maps help in the study of the error in the sinc function methodology. In particular, two domains, D_1 and

D_2, are called *conformally equivalent* if there exists a conformal one-to-one map ϕ of D_1 onto D_2. Denote the inverse mapping (which is also conformal and one-to-one) by ψ. If F is analytic in D_1, then the function $f = (F \circ \psi)\psi'$ is analytic in D_2. If a numerical process such as interpolation has been developed for functions f analytic in D_2, then this process can be carried over to functions F analytic in D_1 via the map ψ.

1.3 Fourier Series and Transforms

There are a few ideas from Fourier series and Fourier transforms that are fundamental to the discussion of various exact properties of the sinc function defined in (1.3). These are collected in this section with the dual purpose of exposing some of the analytic properties of the sinc function and setting the notation used for the spaces in the remainder of the text. The classic Lebesgue p-spaces require the almost everywhere (abbreviation "(a.e.)") qualifier in their definition. This qualifier is included for the sake of accuracy, but the functions dealt with in the text are continuous or at least piecewise continuous.

Definition 1.17 The functions f defined on (a,b) (a.e.) satisfying

$$\|f\|_p \equiv \left\{ \int_a^b |f(x)|^p dx \right\}^{1/p} < \infty, \quad p \geq 1 \qquad (1.9)$$

form a linear space denoted by $L^p(a,b)$. The nonnegative quantity in (1.9) is called the *norm* of f and satisfies

(i) $\|f\|_p = 0$ if and only if $f = 0$,

(ii) $\|\alpha f\|_p = |\alpha| \|f\|_p$ for all $\alpha \in \mathbb{C}$,

(iii) $\|f + g\|_p \leq \|f\|_p + \|g\|_p$.

The distance between two points f and g in $L^p(a,b)$ is measured by the norm of the difference between them. Thus

$$\|f - g\|_p = \left\{ \int_a^b |f(x) - g(x)|^p dx \right\}^{1/p}.$$

The space $L^p(a,b)$ is *complete* (every Cauchy sequence converges in $L^p(a,b)$). That is, if $\{f_n\}$ is a Cauchy sequence in $L^p(a,b)$, then there is a point f in $L^p(a,b)$ such that

$$\|f_n - f\|_p \to 0.$$

If f is in $L^p(a,b)$ and g is in $L^q(a,b)$, where $p+q = pq$, the *Hölder inequality* reads

$$\int_a^b |f(x)g(x)|dx \le \|f\|_p \|g\|_q \qquad (1.10)$$

and if $p = q = 2$, the inequality is called the *Cauchy–Schwarz inequality*

$$\int_a^b |f(x)g(x)|dx \le \|f\|_2 \|g\|_2. \qquad (1.11)$$

In the special case that the interval is the whole real line, $a = -\infty$ and $b = \infty$, the space $L^p(-\infty, \infty)$ will be denoted by $L^p(\mathbb{R})$.

Definition 1.18 Let H be a linear space (vector space). A function $(\cdot, \cdot)$ from $H \times H$ to $\mathbb{C}$ is an *inner product* if the following hold for all f, g, and k in H.

(i) $(g, f) = \overline{(f, g)}$

(ii) $(f + g, k) = (f, k) + (g, k)$

(iii) $(\alpha f, g) = \alpha(f, g)$ for all $\alpha \in \mathbb{C}$

(iv) $(f, f) \ge 0$ with equality if and only if $f \equiv 0$

Definition 1.19 If a linear space H has defined on it an inner product $(\cdot, \cdot)$ and H is complete with respect to the norm defined by

$$\|f\| = (f, f)^{1/2}$$

then H is called a *Hilbert space*.

The cases of most importance for the topics of later chapters are $L^p(a,b)$, where $p = 1, 2$, and ∞. The space $L^2(a,b)$ is a Hilbert space and is distinguished from the other L^p-spaces in that the norm (1.9) (for $p = 2$) arises from the inner product

$$(f, g) = \int_a^b f(x)\overline{g(x)}\,dx.$$

There are distinguished sets in the Hilbert spaces discussed in this section whose properties require the following definition.

Definition 1.20 A set $\{\xi_k\}_{k=1}^{\infty}$ contained in a Hilbert space H is called *orthonormal* if

$$(\xi_k, \xi_n) = \begin{cases} 1 \text{ if } k = n, \\ 0 \text{ if } k \neq n. \end{cases}$$

It is assumed that there is in H a *complete orthonormal set* or an *orthonormal basis*. Thus for every $f \in H$ the sequence of partial sums

$$S_N = \sum_{k=1}^{N}(f, \xi_k)\xi_k$$

converge in $L^2(a, b)$ to f. In other words,

$$\|S_N - f\|_2 \to 0.$$

Parseval's identity distinguishes complete orthonormal sets.

Theorem 1.21 *An orthonormal set $\{\xi_k\}_{k=1}^{\infty}$ contained in a Hilbert space H is a complete orthonormal set if and only if for all $f, g \in H$*

$$(f, g) = \sum_{k=1}^{\infty}(f, \xi_k)(g, \xi_k). \tag{1.12}$$

Setting $f = g$ in Parseval's identity (1.12) gives the equivalent completeness condition that for all $f \in H$

$$\|f\|^2 = \sum_{k=1}^{\infty}|(f, \xi_k)|^2. \tag{1.13}$$

∎

The numbers $\beta_k \equiv (f, \xi_k)$ are called the *Fourier coefficients* of f relative to the orthonormal set $\{\xi_k\}_{k=1}^{\infty}$. The terminology is historical, as indicated in the following example, where the classical Fourier series in $L^2(-P/2, P/2)$ is discussed.

Example 1.22 If a function f is periodic with period P ($f(t+P) = f(t)$ for all $t \in \mathbb{R}$) and $f \in L^2(-P/2, P/2)$, then the Fourier series of f relative to $\{\xi_k(t)\}_{k=-\infty}^{\infty} = \{\exp(2\pi i k t / P)/\sqrt{P}\}_{k=-\infty}^{\infty}$ is

$$f(t) = \sum_{k=-\infty}^{\infty} a_k \exp(2\pi i k t / P)$$

where the Fourier coefficients are

$$a_k = \frac{\beta_k}{\sqrt{P}} = \frac{(f, \xi_k)}{\sqrt{P}} = \frac{1}{P}\int_{-P/2}^{P/2} f(t) \exp(-2\pi i k t / P) dt.$$

The set $\{\exp(2\pi i k t / P)/\sqrt{P}\}_{k=-\infty}^{\infty}$ is a complete orthonormal set in $L^2(-P/2, P/2)$. As a specific case, consider for each fixed $z \in \mathbb{C}$ and $t \in \left(\frac{-\pi}{h}, \frac{\pi}{h}\right)$, $h > 0$ the function $f(t) = \exp(-izt)$. Extend the function by the definition $f(t+P) = f(t)$ so that the function has period $P = 2\pi/h$. The extended function has Fourier coefficients given by

$$\begin{aligned} a_k &= \frac{h}{2\pi} \int_{-\frac{\pi}{h}}^{\frac{\pi}{h}} \exp(-izt) \exp(-ikht) dt \\ &= \frac{(-1)^k h \sin\left(\frac{\pi z}{h}\right)}{\pi(z + kh)} \\ &= \mathrm{sinc}\left(\frac{z + kh}{h}\right). \end{aligned}$$

Hence the Fourier expansion of $f(t) = \exp(-izt)$ is

$$\exp(-izt) = \sum_{k=-\infty}^{\infty} \exp(ikht) \, \mathrm{sinc}\left(\frac{z + kh}{h}\right). \tag{1.14}$$

Since f is piecewise differentiable, this Fourier series converges absolutely and uniformly.

■

If the interval (a,b) is finite and $f \in L^1(a,b)$, then $f \in L^2(a,b)$ by the Cauchy–Schwarz inequality (1.11). This is not the case in $L^1(\mathbb{R})$. If $f \in L^1(\mathbb{R})$, then the *Fourier transform* of f is defined everywhere by the integral

$$\mathcal{F}(f)(x) = \int_{-\infty}^{\infty} f(t)\exp(ixt)dt. \qquad (1.15)$$

If $f \in L^2(\mathbb{R})$, then there are technical details in defining the Fourier transform by the integral in (1.15). In general, the integral has to be replaced by a limit and the Fourier transform is determined (a.e.). For the purposes intended in the next chapter, the *almost everywhere* will play a minor role that is settled in this chapter. The function $f \in L^2(\mathbb{R})$ can be recovered (a.e.) from $\mathcal{F}(f) \in L^2(\mathbb{R})$ by means of the *inversion formula*

$$f(t) = \frac{1}{2\pi} \int_{-\infty}^{\infty} \mathcal{F}(f)(x)\exp(-ixt)dx. \qquad (1.16)$$

Example 1.23 The reciprocal nature of the transform (1.15) and its inverse (1.16) can be illustrated for $f(t) = \pi\exp(-|t|)$. A short calculation, using the fact that the integrand is even, shows that

$$\mathcal{F}(f)(x) = \int_{-\infty}^{\infty} \pi\exp(-|t|)\exp(ixt)dt = \frac{2\pi}{x^2+1}.$$

Two applications of the Residue Theorem 1.13 show that

$$\int_{-\infty}^{\infty} \frac{2\pi}{x^2+1}\exp(-ixt)dx = 2\pi(\pi\exp(-|t|)) = 2\pi f(t).$$

For example, if $t < 0$ use the contour $\gamma = [-r,r] \cup \gamma_r$ $(r > 1)$, where

$$\gamma_r = \{z \in \mathbb{C} : z = re^{i\theta},\ 0 \le \theta \le \pi\}.$$

The second calculation recovered the function f whose Fourier transform was computed in the first calculation. That is, the inversion in (1.16) has been verified for this function. ∎

Example 1.24 Let $f(x) = \text{sinc}(x)$ be defined as in (1.3). Then $f \in L^2(\mathbb{R})$, but $f \notin L^1(\mathbb{R})$. To see that $f \in L^2(\mathbb{R})$, note that

$$\begin{aligned}
\int_{-\infty}^{\infty} |\text{sinc}(x)|^2 dx &= 2\int_0^{\infty} |\text{sinc}(x)|^2 dx \\
&= 2\int_0^1 \text{sinc}^2(x) dx + 2\int_1^{\infty} \text{sinc}^2(x) dx \\
&\leq 2\int_0^1 dx + 2\int_1^{\infty} \frac{dx}{(\pi x)^2} \\
&= 2\left(1 + \frac{1}{\pi^2}\right) < \infty.
\end{aligned}$$

Actually contour integration can be used to show

$$\int_{-\infty}^{\infty} |\text{sinc}(x)|^2 dx = 1.$$

To see that $f \notin L^1(\mathbb{R})$, inscribe triangles under $|\text{sinc}(x)|$ centered at $\left\{\frac{2k-1}{2}\right\}_{k=1}^{\infty}$. This yields

$$\begin{aligned}
\int_{-\infty}^{\infty} |\text{sinc}(x)| dx &= 2\int_0^{\infty} |\text{sinc}(x)| dx \\
&\geq 2\sum_{k=1}^{\infty} \frac{1}{2} \left|\text{sinc}\left(\frac{2k-1}{2}\right)\right| \\
&= \frac{2}{\pi} \sum_{k=1}^{\infty} \frac{1}{2k-1} = \infty.
\end{aligned}$$

∎

A direct method for calculating

$$\mathcal{F}(\text{sinc}(t/h))(x) = \int_{-\infty}^{\infty} \text{sinc}(t/h) \exp(ixt) dt \qquad (1.17)$$

can be based on the Residue Theorem 1.13. The Fourier inversion also provides a convenient method for this calculation. In this direction, for $h > 0$ let $\chi_{(-\pi/h, \pi/h)}$ denote the characteristic function of the interval $(-\pi/h, \pi/h)$ and define

$$\tilde{\chi}_{[-\pi/h, \pi/h]}(x) \equiv \begin{cases} 1/2, & |x| = \pi/h \\ \chi_{(-\pi/h, \pi/h)}, & x \in \mathbb{R} \setminus \{\pm \pi/h\}. \end{cases}$$

PRELIMINARY MATERIAL   15

Integrating $\tilde{\chi}_{[-\pi/h,\pi/h]}(x)$ gives

$$\frac{1}{2\pi}\int_{-\infty}^{\infty}\tilde{\chi}_{[-\pi/h,\pi/h]}(x)\exp(-ixt)dx = (1/h)\operatorname{sinc}(t/h) \qquad (1.18)$$

so that the integral in (1.15) establishes the identity

$$\int_{-\infty}^{\infty}\operatorname{sinc}(t/h)\exp(ixt)dt = h\chi_{(-\pi/h,\pi/h)}(x). \qquad (1.19)$$

The values of $\mathcal{F}(\operatorname{sinc}(t/h)(\pm\pi/h))$ must (if wanted) be determined by separate calculation. The values listed above for $\tilde{\chi}_{[-\pi/h,\pi/h]}$ are the correct ones, that is,

$$\int_{-\infty}^{\infty}\operatorname{sinc}(t/h)\exp(\pm it\pi/h)dt = h/2.$$

A frequently used formula is obtained from (1.19) by putting $t = \xi - u$, where $u \in \mathbb{C}$, and rearranging terms to find

$$\int_{-\infty}^{\infty}\operatorname{sinc}\left(\frac{\xi-u}{h}\right)\exp(ix\xi)d\xi = he^{ixu}\tilde{\chi}_{[-\pi/h,\pi/h]}(x). \qquad (1.20)$$

The fact that the set $\left\{\frac{1}{\sqrt{h}}\operatorname{sinc}\left(\frac{t-kh}{h}\right)\right\}_{k=-\infty}^{\infty}$ is an orthonormal set is a consequence of (1.20) and the following integral analogue of Parseval's identity (1.12).

Theorem 1.25 (Parseval's Theorem) *If f and g are in $L^2(\mathbb{R})$, then*

$$\int_{-\infty}^{\infty} f(t)\overline{g(t)}\,dt = \frac{1}{2\pi}\int_{-\infty}^{\infty}\mathcal{F}(f)(x)\overline{\mathcal{F}(g)(x)}\,dx.$$

∎

Example 1.26 Applying Parseval's Theorem 1.25 with

$$f(t) = \operatorname{sinc}\left(\frac{t-kh}{h}\right)$$

and

$$\overline{g(t)} = \operatorname{sinc}\left(\frac{t-nh}{h}\right)$$

results in the identity

$$\int_{-\infty}^{\infty} \text{sinc}\left(\frac{t-kh}{h}\right) \text{sinc}\left(\frac{t-nh}{h}\right) dt$$
$$= \frac{1}{2\pi} \int_{-\infty}^{\infty} \mathcal{F}(f)(x)\overline{\mathcal{F}(g)(x)}\, dx$$
$$= \frac{h^2}{2\pi} \int_{-\infty}^{\infty} e^{ikhx} e^{-inhx} \tilde{\chi}_{[-\pi/h,\pi/h]}(x)\, dx,$$

where (1.20) has been used to obtain the last equality. Thus

$$\int_{-\infty}^{\infty} \text{sinc}\left(\frac{t-kh}{h}\right) \text{sinc}\left(\frac{t-nh}{h}\right) dt = h\delta_{k,n}^{(0)} \qquad (1.21)$$

where the Kronecker delta is denoted by

$$\delta_{k,n}^{(0)} \equiv \begin{cases} 0 & n \neq k, \\ 1 & n = k. \end{cases}$$

Hence the set $\left\{\frac{1}{\sqrt{h}} \text{sinc}\left(\frac{t-kh}{h}\right)\right\}_{k=-\infty}^{\infty}$ is an orthonormal set.

∎

If $\mathcal{F}(f)$ is in $L^2(-\pi/h, \pi/h)$ and

$$f(z) = \frac{1}{2\pi} \int_{-\frac{\pi}{h}}^{\frac{\pi}{h}} \mathcal{F}(f)(x) \exp(-izx) dx, \qquad (1.22)$$

then Morera's Theorem 1.6 shows that f is an entire function. Taking absolute values in (1.22) yields the inequality

$$|f(z)| \leq \exp(\pi|z|/h) \frac{1}{2\pi} \int_{-\frac{\pi}{h}}^{\frac{\pi}{h}} |\mathcal{F}(f)(x)| dx \leq K \exp(\pi|z|/h). \quad (1.23)$$

Entire functions satisfying this growth constraint are said to be of *exponential type* π/h. Thus, functions f defined by (1.22) have the following properties: they are entire, are of exponential type π/h, and when restricted to $\mathbb{R}$ (by employing (1.16)), they are in $L^2(\mathbb{R})$. This is illustrated in (1.18) for $\mathcal{F}(f)(x) = \tilde{\chi}_{[-\pi/h,\pi/h]}(x)$. That is, the sinc function has all of these properties. Exercise 1.3 shows that for the sinc function the value of K may be taken to be one.

The identity in (1.19) shows that the Fourier transform of the sinc function is compactly supported. One is led to wonder whether this situation is special to the sinc function. In other words, if a function f is entire, satisfies (1.23), and belongs to $L^2(\mathbb{R})$, then is it the case that f has a compactly supported Fourier transform? This converse is the content of the very important Paley–Wiener Theorem.

Theorem 1.27 (Paley–Wiener Theorem) *Assume that f is entire and $f \in L^2(\mathbb{R})$. If there are positive constants K and π/h so that for all $z \in \mathbb{C}$*

$$|f(z)| \leq K \exp(\pi |z|/h), \tag{1.24}$$

then $\mathcal{F}(f) \in L^2(-\pi/h, \pi/h)$ and

$$f(z) = \frac{1}{2\pi} \int_{-\frac{\pi}{h}}^{\frac{\pi}{h}} \mathcal{F}(f)(x) \exp(-ixz) dx. \tag{1.25}$$

■

Exercises

Exercise 1.1 Find the image of the disc

$$D = \{z \in \mathbb{C} : |z - 1/2| < 1/2\}$$

under the linear fractional transformation

$$L(z) = \frac{z}{1-z}.$$

It helps to notice that the real line is mapped to the real line by L with the interval $(0,1)$ going to $(0,\infty)$. Parameterize the upper arc of the circle by $z(\theta) = (1 + \exp(i\theta))/2, 0 \leq \theta \leq \pi$ and show that this is mapped to the positive imaginary axis. The region bounded by the upper arc of the circle and the real line is mapped to the "positive" quarter plane. For $0 < d \leq \frac{\pi}{2}$, "clamp" the disc D by the restriction

$$D_E = \left\{ z \in \mathbb{C} : \left| \arg\left(\frac{z}{1-z}\right) \right| < d \right\}$$

and find the image of D_E under L. Note that for $d = \pi/2$, D_E is the domain D. Find the image of $L(D_E)$ under the map $\ell n(z)$ in Example 1.16.

Exercise 1.2 The calculation of the integral

$$I(k) = \int_{-\infty}^{\infty} \frac{\text{sinc}\left(\frac{t-kh}{h}\right)}{t^2+1} dt$$

could be dealt with directly, for example, by using the Residue Theorem 1.13. However, using (1.20) along with the results of Example 1.23 and Parseval's Theorem 1.25 shows that

$$I(k) = \frac{h}{2}\int_{-\infty}^{\infty} e^{-|x|} e^{-ixkh} \tilde{\chi}_{[-\pi/h,\pi/h]}(x)dx.$$

Direct integration of the latter gives

$$I(k) = h\left[\frac{1+(-1)^{k+1}\exp(-\pi/h)}{(kh)^2+1}\right].$$

Exercise 1.3 By Example 1.10 the function $\text{sinc}(z/h)$, $h > 0$, is entire. Show that it satisfies

$$|\text{sinc}(z/h)| \leq 1 \exp((\pi/h)|z|)$$

using the integral representation of $\text{sinc}(z/h)$ found in (1.18). Thus, for this function, $K = 1$ in (1.24).

Exercise 1.4 An entire function that is in $L^2(\mathbb{R})$ is given by

$$f(z) = \exp(-z^2).$$

Calculate the Fourier transform of f using the fact that the given function is even and the identity

$$\int_{-\infty}^{\infty} \exp(-t^2)\cos(xt)dt = \sqrt{\pi}\exp(-(x^2)/4)$$

which may be found in [2, page 480]. Note that $f(z) = \exp(-z^2)$ does not satisfy (1.24).

Exercise 1.5 For a function f defined on all of $\mathbb{R}$ consider the formal series

$$f(t) = \sum_{k=-\infty}^{\infty} a_k \text{ sinc}\left[(t-kh)/h\right].$$

Multiply this equation by $\text{sinc}\left[(t-nh)/h\right]$ and integrate the result over $\mathbb{R}$ to formally find the values of the a_k. Remember the relation in (1.21).

References

[1] L. V. Ahlfors, *Complex Analysis*, Third Ed., McGraw-Hill, Inc., New York, 1979.

[2] I. S. Gradshteyn and I. M. Ryzhik, *Table of Integrals, Series and Products*, Academic Press, Inc., New York, 1980.

[3] E. Hille, *Analytic Function Theory*, Vol. II, Second Ed., Chelsea Publishing Co., New York, 1977.

[4] J. E. Marsden and M. J. Hoffman, *Basic Complex Analysis*, Second Ed., W. H. Freeman and Co., New York, 1987.

[5] W. Rudin, *Real and Complex Analysis*, Third Ed., McGraw-Hill, Inc., New York, 1987.

[6] E. C. Titchmarsh, *Theory of Fourier Integrals*, Clarendon Press, Oxford, 1954.

Chapter 2

Numerical Methods on the Real Line

The theory of sinc series (cardinal functions) on the whole real line is developed in this chapter. The material in Section 2.1 for the Paley–Wiener class of functions provides formulas where sinc interpolation and quadrature are exact. Section 2.2 discusses the more practical use of sinc series as a tool for approximating functions in a class that is less restrictive than is the Paley–Wiener class. The error in these infinite sample approximations is developed via contour integration. With the introduction of appropriate growth restrictions of the functions on the real line, a subclass is identified for which these formulas are useful as practical numerical tools. Examples are used to motivate the restrictions and advertise future topics in the text.

2.1 Exact Interpolation and Quadrature

The remarkable fact that a function interpolating f at a countable collection of points on the real line can, for the appropriate class of functions f, be exact for all $z \in \mathbb{C}$ is the foundation of this section. This leads, via the orthogonality exhibited in (1.21), to an exact quadrature rule on $\mathbb{R}$ for the integrable functions in this class. The definition of the cardinal function, which is an infinite series involving sinc basis functions, will begin the development.

Definition 2.1 Let f be a function defined on $\mathbb{R}$ and let $h > 0$. Define the series

$$C(f,h)(x) \equiv \sum_{k=-\infty}^{\infty} f(kh)\,\text{sinc}\left(\frac{x-kh}{h}\right) \qquad (2.1)$$

where from (1.3)

$$\text{sinc}\left(\frac{x-kh}{h}\right) \equiv \begin{cases} \frac{\sin(\pi(x-kh)/h)}{\pi(x-kh)/h} &, \; x \neq kh \\ 1 &, \; x = kh. \end{cases} \qquad (2.2)$$

Whenever the series in (2.1) converges it is called the *cardinal function* of f.

It is easy to see that the cardinal function interpolates f at the points $\{nh\}_{n=-\infty}^{\infty}$. This series was first discussed in [8], and in a slightly different form it was studied extensively in [9]. A class of functions where the cardinal function of f converges to f is characterized in the following definition.

Definition 2.2 Let h be a positive constant. The *Paley–Wiener class* of functions $B(h)$ is the family of entire functions f such that on the real line $f \in L^2(\mathbb{R})$ and in the complex plane f is of exponential type π/h, i.e.,

$$|f(z)| \leq K \exp(\pi|z|/h)$$

for some $K > 0$.

Several properties of the Paley–Wiener class of functions $B(h)$ were developed in [6] and [7]. The following theorems help characterize this class.

Theorem 2.3 *If $f \in B(h)$, then for all $z \in \mathbb{C}$,*

$$f(z) = \frac{1}{h} \int_{-\infty}^{\infty} f(t) \operatorname{sinc}\left(\frac{t-z}{h}\right) dt. \tag{2.3}$$

Proof The Paley–Wiener Theorem 1.27 shows that

$$\begin{aligned}
f(z) &= \frac{1}{2\pi} \int_{-\pi/h}^{\pi/h} \mathcal{F}(f)(x) e^{-ixz} dx \\
&= \frac{1}{2\pi} \int_{-\pi/h}^{\pi/h} \left[\int_{-\infty}^{\infty} f(t) e^{ixt} dt \right] e^{-ixz} dx \\
&= \frac{1}{2\pi} \int_{-\infty}^{\infty} f(t) \left[\int_{-\pi/h}^{\pi/h} e^{ix(t-z)} dx \right] dt \\
&= \frac{1}{2\pi} \int_{-\infty}^{\infty} f(t) \left[\frac{e^{i\pi(t-z)/h} - e^{-i\pi(t-z)/h}}{i(t-z)} \right] dt \\
&= \frac{1}{h} \int_{-\infty}^{\infty} f(t) \operatorname{sinc}\left(\frac{t-z}{h}\right) dt.
\end{aligned}$$

∎

From Exercise 1.3, $\operatorname{sinc}(z/h)$ belongs to $B(h)$. Further examples of functions in $B(h)$ may be constructed using the following theorem.

Theorem 2.4 *If $g \in L^2(\mathbb{R})$, then $k \in B(h)$, where*

$$k(z) \equiv \frac{1}{h} \int_{-\infty}^{\infty} g(t) \operatorname{sinc}\left(\frac{t-z}{h}\right) dt. \tag{2.4}$$

Proof Apply Parseval's Theorem 1.25 to k and use (1.20) to find

$$\begin{aligned}
k(z) &= \frac{1}{2\pi h} \int_{-\infty}^{\infty} \mathcal{F}(g)(x) \mathcal{F}\left(\operatorname{sinc}\left(\frac{t-z}{h}\right)\right)(-x) dx \\
&= \frac{1}{2\pi} \int_{-\pi/h}^{\pi/h} \mathcal{F}(g)(x) e^{-ixz} dx.
\end{aligned}$$

This representation, along with the discussion preceding the Paley–Wiener Theorem 1.27, implies that $k \in L^2(\mathbb{R})$ and is entire.

∎

The following theorem and corollary give the exact interpolation and quadrature formulas for functions in $B(h)$.

Theorem 2.5 *If $f \in B(h)$, then for all $z \in \mathbb{C}$*

$$f(z) = \sum_{k=-\infty}^{\infty} f(kh) \operatorname{sinc}\left(\frac{z-kh}{h}\right), \tag{2.5}$$

and

$$f(kh) = \frac{1}{h} \int_{-\infty}^{\infty} f(t) \operatorname{sinc}\left(\frac{t-kh}{h}\right) dt. \tag{2.6}$$

Proof Since $f \in B(h)$ the Paley–Wiener Theorem 1.27 gives

$$f(z) = \frac{1}{2\pi} \int_{-\pi/h}^{\pi/h} \mathcal{F}(f)(x) e^{-ixz} dx.$$

Substituting the series expansion (1.14) in the integral on the right-hand side yields

$$\begin{aligned} f(z) &= \frac{h}{\pi} \sin\left[\frac{\pi z}{h}\right] \sum_{k=-\infty}^{\infty} \frac{(-1)^k}{(z+kh)} \left[\frac{1}{2\pi} \int_{-\pi/h}^{\pi/h} \mathcal{F}(f)(x) e^{ikhx} dx\right] \\ &= \sum_{k=-\infty}^{\infty} \frac{\sin[\pi(z+kh)/h]}{\pi(z+kh)/h} f(-kh) \\ &= \sum_{k=-\infty}^{\infty} f(kh) \operatorname{sinc}\left(\frac{z-kh}{h}\right). \end{aligned}$$

The interchange of the sum and the integral is justified by the uniform convergence of the series in (1.14). The coefficients $f(-kh)$ occurring in the second equality follow from the evaluation of the integral using the Paley–Wiener Theorem 1.27. Substituting $z = kh$ in Theorem 2.3 yields (2.6).

∎

Corollary 2.6 *If $f \in B(h)$ and $f \in L^1(\mathbb{R})$, then*

$$\int_{-\infty}^{\infty} f(t) dt = h \sum_{k=-\infty}^{\infty} f(kh). \tag{2.7}$$

Proof Replace z by $t \in \mathbb{R}$ in (2.5) and integrate the result over $\mathbb{R}$. Since $f \in L^1(\mathbb{R})$, the series may be integrated term by term and the integral

$$\int_{-\infty}^{\infty} \operatorname{sinc}\left(\frac{t-kh}{h}\right) dt = \int_{-\infty}^{\infty} \operatorname{sinc}\left(\frac{s}{h}\right) ds = h$$

is evaluated by using (1.20) with $x = 0$. ∎

If $f \in B(h)$, then the Fourier series of f relative to the set $\left\{\frac{1}{\sqrt{h}} \operatorname{sinc}\left(\frac{t-kh}{h}\right)\right\}_{k=-\infty}^{\infty}$ is

$$\sum_{k=-\infty}^{\infty} \frac{\beta_k}{\sqrt{h}} \operatorname{sinc}\left(\frac{t-kh}{h}\right)$$

where from Theorem 2.5

$$\beta_k = \sqrt{h}\, f(kh).$$

This set has the following properties in $B(h)$.

Theorem 2.7 *If $f \in B(h)$, then*

$$\|f\|_2^2 \equiv \int_{-\infty}^{\infty} |f(t)|^2 dt = h \sum_{k=-\infty}^{\infty} |f(kh)|^2 = \sum_{k=-\infty}^{\infty} |\beta_k|^2.$$

Thus the set $\left\{\frac{1}{\sqrt{h}} \operatorname{sinc}\left(\frac{t-kh}{h}\right)\right\}_{k=-\infty}^{\infty}$ is a complete orthonormal set in $B(h)$.

Proof The set is orthonormal by (1.21). To show that it is complete, use the representation of f in (2.5) to find

$$\int_{-\infty}^{\infty} |f(t)|^2 dt = \int_{-\infty}^{\infty} f(t)\overline{f(t)} dt$$

$$= \int_{-\infty}^{\infty} \left[\sum_{k=-\infty}^{\infty} \frac{\beta_k}{\sqrt{h}} \operatorname{sinc}\left(\frac{t-kh}{h}\right)\right] \left[\sum_{\ell=-\infty}^{\infty} \frac{\overline{\beta_\ell}}{\sqrt{h}} \operatorname{sinc}\left(\frac{t-\ell h}{h}\right)\right] dt$$

$$= \sum_{k=-\infty}^{\infty} \sum_{\ell=-\infty}^{\infty} \beta_k \overline{\beta_\ell} \left[\frac{1}{h} \int_{-\infty}^{\infty} \operatorname{sinc}\left(\frac{t-kh}{h}\right) \operatorname{sinc}\left(\frac{t-\ell h}{h}\right) dt\right]$$

$$= \sum_{k=-\infty}^{\infty} |\beta_k|^2$$

$$= h \sum_{k=-\infty}^{\infty} |f(kh)|^2.$$

The last two equalities follow from Exercise 1.5 and (2.6), respectively. Hence the set is complete by Parseval's identity (1.13).

∎

It is convenient here to collect the two exact relationships on interpolation and quadrature for functions $f \in B(h)$. They use the translated sinc functions, (2.9), three representatives of which are graphically illustrated in Figure 2.1.

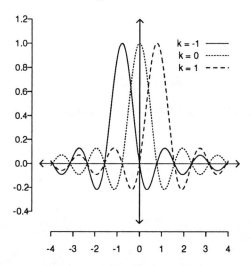

Figure 2.1 Three members $(S(k,h)(x), k = -1, 0, 1, h = \pi/4)$ of the sinc basis.

Theorem 2.8 *If $f \in B(h)$, then for all $z \in \mathbb{C}$,*

$$f(z) = C(f,h)(z) = \sum_{k=-\infty}^{\infty} f(kh) S(k,h)(z) \qquad (2.8)$$

where the translated sinc functions are given by

$$S(k,h)(z) \equiv \operatorname{sinc}\left(\frac{z - kh}{h}\right). \qquad (2.9)$$

Furthermore, if $f \in L^1(\mathbb{R})$,

$$\int_{-\infty}^{\infty} f(t)dt = h \sum_{k=-\infty}^{\infty} f(kh). \tag{2.10}$$

∎

The translated sinc functions defined in (2.9) will be used extensively in the numerical solution of differential equations. The second identity in Theorem 2.8 is the *trapezoidal rule* of integration.

The applicability of the sinc function basis in the processes of interpolation and numerical integration for functions that may not be members of $B(h)$ is explored in the following three examples.

Example 2.9 The solution of the one-dimensional heat equation

$$\begin{aligned} u_t(x,t) &= u_{xx}(x,t), \quad x \in \mathbb{R}, \quad t > 0 \\ u(x,0) &= f(x), \quad x \in \mathbb{R} \end{aligned}$$

for f bounded and continuous on $\mathbb{R}$ is given by

$$u(x,t) = \frac{1}{\sqrt{4\pi t}} \int_{-\infty}^{\infty} e^{-(\xi-x)^2/(4t)} f(\xi) d\xi. \tag{2.11}$$

A problem encountered in discrete observability is the construction of the initial data f from discrete sampling of the solution u. One technique, using a cardinal function expansion of f, will be outlined here. Further details on this technique are in [2] and [4]. Assume that $f \in B(h)$ so that sinc interpolation is exact, ((2.8) holds on all of $\mathbb{R}$). Substitution of the sinc expansion of f into (2.11) yields

$$u(x,t) = \frac{1}{\sqrt{4\pi t}} \sum_{k=-\infty}^{\infty} f(kh) \int_{-\infty}^{\infty} e^{-(\xi-x)^2/(4t)} \operatorname{sinc}\left(\frac{\xi - kh}{h}\right) d\xi.$$

Evaluating this at the node $x_n = nh$ and changing variables via $s = (\xi - nh)/h$ leads to

$$u(x_n,t) = \frac{h}{\sqrt{4\pi t}} \sum_{k=-\infty}^{\infty} f(kh) \int_{-\infty}^{\infty} e^{-s^2 h^2/(4t)} \operatorname{sinc}(s - \ell) ds \tag{2.12}$$

where $\ell = k - n$. Using the result

$$\text{sinc}(s - \ell) = \frac{1}{2\pi} \int_{-\pi}^{\pi} e^{-is\tau} e^{i\ell\tau} d\tau$$

and the identity

$$\int_{-\infty}^{\infty} \exp\left(-ivw - \frac{w^2}{4t}\right) dw = 2\sqrt{\pi t} \, \exp(-v^2 t),$$

the right-hand side of (2.12) can be put into the form

$$u(x_n, t) = \frac{1}{2\pi} \sum_{k=-\infty}^{\infty} f(kh) \int_{-\pi}^{\pi} e^{-\tau^2 t/h^2} e^{i\ell\tau} d\tau.$$

Replace $f(kh)$ with f_k and regard the f_k as unknown coefficients of a sinc expansion of the initial heat ((2.8) with $f(kh)$ replaced by f_k) to find

$$u(x_n, t) \simeq \frac{1}{2\pi} \sum_{k=-M}^{M} f_k \int_{-\pi}^{\pi} e^{-\tau^2 t/h^2} e^{i\ell\tau} d\tau, \quad n = -M, \ldots, M.$$

This provides an approximation for $\{f_k\}_{k=-M}^{M}$ at any time t. Sampling $u(x_n, t)$ at $\hat{t} = (h/(2\pi))^2$ and remembering $\ell = k - n$ yields

$$u(x_n, \hat{t}) = \sum_{k=-M}^{M} f_k \beta_{k-n}, \quad n = -M, \ldots, M$$

where

$$\beta_{k-n} \equiv \frac{1}{2\pi} \int_{-\pi}^{\pi} e^{-\tau^2/(4\pi^2)} e^{i(k-n)\tau} d\tau.$$

This may be written as the linear system

$$B\vec{f} = \vec{u} \tag{2.13}$$

where B is the $m \times m$ ($m = 2M + 1$) symmetric centrosymmetric Toeplitz matrix with entries $b_{n,k} = \beta_{k-n}$ in Definition A.15.

∎

The exact relationship in (2.12) depends on the initial data f belonging to $B(h)$. The next example explores the consequences of interpolating a function f that does not belong to $B(h)$ with its cardinal function (2.1).

Example 2.10 Associate with the function $f(t) = (t^2+1)^{-1}$ the series

$$f(t) \approx \sum_{k=-\infty}^{\infty} a_k \operatorname{sinc}\left(\frac{t-kh}{h}\right). \tag{2.14}$$

Proceeding formally (acting as if $f \in B(h)$), the coefficients a_k are given in Theorem 2.5 as

$$\begin{aligned} a_k &= \frac{1}{h} \int_{-\infty}^{\infty} \frac{1}{t^2+1} \operatorname{sinc}\left(\frac{t-kh}{h}\right) dt \\ &= \frac{1}{2\pi h} \int_{-\infty}^{\infty} \pi e^{-|x|} \, \overline{he^{ixkh}\tilde{\chi}_{[-\pi/h,\pi/h]}(x)} \, dx. \end{aligned}$$

The calculation from Exercise 1.2 shows that

$$\begin{aligned} a_k &= \frac{1}{k^2h^2+1} + (-1)^{k+1}\frac{e^{-\pi/h}}{k^2h^2+1} \\ &= f(kh)\left[1 + (-1)^{k+1}e^{-\pi/h}\right]. \end{aligned}$$

Hence the formal series in (2.14) is not the cardinal function for f since $a_k \neq f(kh)$. Substituting these a_k into (2.14) and continuing the formalism by using (2.5) yields

$$\begin{aligned} \epsilon(t) &\equiv f(t) - \sum_{k=-\infty}^{\infty} a_k \operatorname{sinc}\left(\frac{t-kh}{h}\right) \\ &\qquad\qquad\qquad\qquad\qquad\qquad\qquad\qquad (2.15) \\ &= -e^{-\pi/h} \sum_{k=-\infty}^{\infty} \frac{(-1)^{k+1} \operatorname{sinc}((t-kh)/h)}{k^2h^2+1}. \end{aligned}$$

It can be shown (Exercise 2.3) that

$$f(t) - C(f,h)(t) = \frac{-1}{t^2+1}\left[\frac{t\sin(\pi t/h)}{\sinh(\pi/h)}\right] \tag{2.16}$$

and this is equal to the series on the right-hand side of (2.15). Hence $\epsilon(t) = f(t) - C(f,h)(t)$. From (2.16) one finds

$$|\epsilon(t)| \leq \frac{1}{\sinh(\pi/h)} = \begin{cases} .366 \times 10^{-1}, & h = \pi/4 \\ .671 \times 10^{-3}, & h = \pi/8 \\ .225 \times 10^{-6}, & h = \pi/16. \end{cases}$$

For $h \leq 2\pi/(\ell n(2))$

$$\frac{1}{\sinh(\pi/h)} = \frac{2}{e^{\pi/h} - e^{-\pi/h}} = \frac{2e^{-\pi/h}}{1 - e^{-2\pi/h}} \leq 4e^{-\pi/h}.$$

Hence the interpolation of $f(t) = (t^2 + 1)^{-1}$ by the formal series in (2.14) has the exponential convergence rate $\mathcal{O}(e^{-\pi/h})$. Thus infinite interpolation of f is extremely good even though $f \notin B(h)$. Note however that for a practical interpolation scheme one needs to consider the error incurred when truncating the series. Thus from (2.15)

$$f(t) - \sum_{k=-M}^{M} a_k \operatorname{sinc}\left(\frac{t - kh}{h}\right)$$

$$= \sum_{|k|=M+1}^{\infty} a_k \operatorname{sinc}\left(\frac{t - kh}{h}\right) + \epsilon(t). \tag{2.17}$$

The error from truncation is bounded by

$$\left| \sum_{|k|=M+1}^{\infty} a_k \operatorname{sinc}\left(\frac{t - kh}{h}\right) \right|$$

$$\leq (2 + 2e^{-\pi/h}) \sum_{k=M+1}^{\infty} |f(kh)|$$

$$= (2 + 2e^{-\pi/h}) \sum_{k=M+1}^{\infty} \frac{1}{k^2 h^2 + 1} \tag{2.18}$$

$$= \mathcal{O}((Mh^2)^{-1}), \quad Mh^2 \to \infty$$

so that the error on the right-hand side of (2.17) is given by the right-hand side of (2.18) in spite of the rapid convergence of $\epsilon(t)$. The contribution to the error from $\epsilon(t)$ (exponential convergence) and the contribution from the truncation error (2.18) (algebraic convergence) are, in a sense, out of balance with regard to their order.

∎

It is clear from the previous example that to achieve an exponential rate of convergence, as a function of the number of points sampled, more restrictive assumptions on the rate of decay of f on the real line are required. One may expect to have more accuracy for a function like $f(t) = \exp(-|t|)$ because in this case the right-hand side of (2.18) is replaced by $\mathcal{O}(\exp(-Mh))$. The following example shows that exponential decay of the function f is not the only criterion required to guarantee an exponential convergence rate.

Example 2.11 Consider the use of the trapezoidal rule (2.10) for the approximation of the integrals of the functions

$$f_1(t) = (\pi/2)\exp(-|t|)$$
$$f_2(t) = \text{sech}(t) = \frac{1}{\cosh(t)}.$$

For $i = 1, 2$ one can write

$$\int_{-\infty}^{\infty} f_i(t)dt = S_{M,i} + T_{M,i} + \eta_i$$

where

$$S_{M,i} \equiv h \sum_{k=-M}^{M} f_i(kh)$$

$$T_{M,i} \equiv h \sum_{|k|=M+1}^{\infty} f_i(kh)$$

and

$$\eta_i \equiv \int_{-\infty}^{\infty} f_i(t)dt - h \sum_{k=-\infty}^{\infty} f_i(kh).$$

Note that neither of the functions f_i belongs to $B(h)$. The results in Table 2.1 show that in the case of f_1 the trapezoidal rule converges with the algebraic rate $\mathcal{O}(h^2)$, whereas the trapezoidal rule applied to f_2 converges with the exponential rate $\mathcal{O}(\exp(-\pi\sqrt{M}))$. This is reported in the last column in Table 2.1 as the numerical order constant (NOC) defined by

$$NOC \equiv \frac{AE \text{ (absolute error)}}{CR \text{ (convergence rate)}} = \frac{|\pi - S_{M,i}|}{CR}.$$

$$\int_{-\infty}^{\infty} \frac{\pi}{2} \exp(-|t|) dt = \pi \approx S_{M,1}$$

M	$h = \pi/\sqrt{M}$	AE	$\frac{AE}{h^2}$
20	.702	$.128 \times 10^{-0}$	.2597
40	.497	$.643 \times 10^{-1}$	.2607
80	.351	$.322 \times 10^{-1}$	.2613
160	.248	$.161 \times 10^{-1}$	.2615
320	.176	$.807 \times 10^{-2}$	.2617

$$\int_{-\infty}^{\infty} \frac{dt}{\cosh(t)} = \pi \approx S_{M,2}$$

M	$h = \pi/\sqrt{M}$	AE	$\frac{AE}{\exp(-\pi\sqrt{M})}$
4	1.571	$.204 \times 10^{-1}$	10.941
8	1.111	$.144 \times 10^{-2}$	10.386
16	.785	$.346 \times 10^{-4}$	9.934
32	.555	$.183 \times 10^{-6}$	9.575
64	.393	$.113 \times 10^{-9}$	9.301

Table 2.1 Algebraic contrasted with exponential convergence for the trapezoidal rule.

The quantity NOC should approach a constant, independent of M, if the correct convergence rate of a method has been established. The convergence rate $\mathcal{O}(\exp(-\pi\sqrt{M}))$ for the second function will be analytically established in Theorem 2.21. In contrast to the interpolation in Example 2.10, both functions in this example go to zero exponentially:

$$f_i(t) = \mathcal{O}(e^{-|t|}), \quad \pm t \to \infty, \quad i = 1, 2.$$

Thus more is needed for the exponential convergence of the truncated trapezoidal rule than exponential decay of f. Note that $f_1(t)$ is not analytic in any domain containing the origin. Exercise 2.1 shows, in a very special case, that exponential convergence can be very rapid.

■

2.2 Approximate Interpolation and Quadrature

Whereas the interpolation and quadrature formulas of Theorem 2.8 are exact, the last two examples of the previous section show that there are functions $f \notin B(h)$ where the approximation of f by its cardinal series (or the approximation of the integral of f by the trapezoidal rule) has errors that decrease exponentially. One is led to seek a less restrictive class of functions than $B(h)$ where the exponential decay of the error can be maintained. Collaterally, the error due to the truncation of the infinite series defining the approximation must be balanced with these exponential errors. Recall that for f to be a member of the Paley–Wiener class $B(h)$ f must be entire, $f \in L^2(\mathbb{R})$, and f must be of exponential type π/h. To assume $f \in L^2(\mathbb{R})$ (or some $L^p(\mathbb{R})$) seems a reasonable requirement on f if one is trying to approximate the integral of f over $\mathbb{R}$. In an attempt to replace the class $B(h)$ by a less restrictive class of functions in which the formulas of Theorem 2.8 are nevertheless "very accurate," one is led to look a little more closely at the assumption that f is entire. Whereas one may relinquish this highly restrictive assumption, one does not want to dispense with the tools of analytic function theory. Since the approximations of f will be on the real line, it seems the assumption that f is analytic in a domain containing $\mathbb{R}$ is a reasonable beginning.

Definition 2.12 Let D_S denote the *infinite strip domain* (Figure 2.2) of width $2d$, $d > 0$, given by

$$D_S \equiv \{z \in \mathbb{C} : z = x + iy, \ |y| < d\}. \qquad (2.19)$$

Let $B^p(D_S)$ be the set of functions analytic in D_S that satisfy

$$\int_{-d}^{d} |f(t+iy)| dy = \mathcal{O}(|t|^a), \quad t \to \pm\infty, \ 0 \leq a < 1 \qquad (2.20)$$

and

$$N^p(f, D_S) \equiv \lim_{y \to d^-} \left\{ \left(\int_{-\infty}^{\infty} |f(t+iy)|^p dt \right)^{1/p} \right.$$

$$\left. + \left(\int_{-\infty}^{\infty} |f(t-iy)|^p dt \right)^{1/p} \right\} < \infty. \qquad (2.21)$$

For $p = 1$, let $N(f, D_S) \equiv N^1(f, D_S)$ and $B(D_S) \equiv B^1(D_S)$.

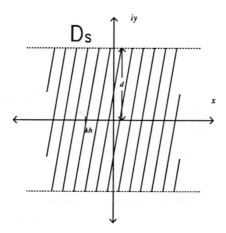

Figure 2.2 The infinite strip domain D_S of total width $2d, d > 0$.

The following theorem gives an interpolation result in $B^p(D_S)$.

Theorem 2.13 *If $f \in B^p(D_S)$ ($p = 1$ or 2) and $h > 0$, then*

$$\begin{aligned} f(x) - C(f,h)(x) &= f(x) - \sum_{k=-\infty}^{\infty} f(kh) \operatorname{sinc}\left(\frac{x-kh}{h}\right) \\ &\equiv \epsilon(x) \qquad (2.22) \\ &= S_h(x) I(f,h)(x). \end{aligned}$$

Here

$$S_h(x) \equiv \frac{\sin\left(\frac{\pi x}{h}\right)}{2\pi i} \qquad (2.23)$$

and

$$I(f,h)(x) \qquad (2.24)$$

$$\equiv \int_{-\infty}^{\infty} \left\{ \frac{F(x, t - id^-)}{\sin(\pi(t - id^-)/h)} - \frac{F(x, t + id^-)}{\sin(\pi(t + id^-)/h)} \right\} dt$$

where

$$F(s, u \pm iv) = \frac{f(u \pm iv)}{(u - s \pm iv)}. \qquad (2.25)$$

Moreover, if $f \in B^p(D_S)$, $p = 1$ or 2, then

$$\|f - C(f,h)\|_\infty \leq \frac{N^p(f, D_S)}{2(\pi d)^{1/p} \sinh(\pi d/h)} = \mathcal{O}(e^{-\pi d/h}). \qquad (2.26)$$

Proof Let $y_n = d - \frac{1}{n}$ and define the domain

$$R_n \equiv \left\{ z \in \mathbb{C} : z = t + iy, |y| < y_n, -\left(n + \frac{1}{2}\right)h < t < \left(n + \frac{1}{2}\right)h \right\}.$$

Let ∂R_n denote the boundary of R_n and note that $R_n \subset D_S$. Applying the Residue Theorem 1.13 to the integral of

$$g(z) = \frac{\sin\left(\frac{\pi x}{h}\right) f(z)}{(z - x) \sin\left(\frac{\pi z}{h}\right)}$$

around ∂R_n yields (see Example 1.14)

$$\begin{aligned} I_n(x) &\equiv S_h(x) \int_{\partial R_n} \frac{f(z) dz}{(z - x) \sin(\pi z/h)} \\ &\qquad\qquad\qquad\qquad\qquad\qquad\qquad\qquad (2.27) \\ &= f(x) - \sum_{k=-n}^{n} f(kh) \operatorname{sinc}\left(\frac{x - kh}{h}\right). \end{aligned}$$

Thus
$$\lim_{n \to \infty} I_n(x) = \epsilon(x).$$

Also, integrating around the boundary of R_n gives

$$I_n(x) = S_h(x) \Bigg\{ \int_{-y_n}^{y_n} \frac{F(x, (n+1/2)h + iy) \, i \, dy}{\sin[\pi((n+1/2)h + iy)/h]}$$
$$+ \int_{y_n}^{-y_n} \frac{F(x, -(n+1/2)h + iy) \, i \, dy}{\sin[\pi(-(n+1/2)h + iy)/h]}$$
$$+ \int_{(n+1/2)h}^{-(n+1/2)h} \frac{F(x, t + iy_n) \, dt}{\sin[\pi(t + iy_n)/h]} \quad (2.28)$$
$$+ \int_{-(n+1/2)h}^{(n+1/2)h} \frac{F(x, t - iy_n) \, dt}{\sin[\pi(t - iy_n)/h]} \Bigg\}.$$

On the vertical segments of ∂R_n, using the fact that $\cos(iy) = \cosh(y)$ and $\sin(iy) = i \sinh(y)$ yields

$$\sin[\pi(\pm(n+1/2)h + iy)/h]$$
$$= \pm \sin\left(\frac{2n+1}{2}\pi\right) \cosh(\pi y/h) + i \cos\left(\frac{2n+1}{2}\pi\right) \sinh(\pi y/h)$$
$$= \pm(-1)^n \cosh(\pi y/h)$$

so that

$$|\sin[\pi(\pm(n+1/2)h + iy)/h]| = \cosh(\pi y/h) \geq 1. \quad (2.29)$$

Also

$$|\pm(n+1/2)h + iy - x| \geq |\pm(n+1/2)h - x|. \quad (2.30)$$

Hence, using (2.29) and (2.30) one finds

$$\left| \int_{-y_n}^{y_n} \frac{F(x, (n+1/2)h + iy) \, i \, dy}{\sin[\pi((n+1/2)h + iy)/h]} \right|$$
$$\leq \frac{1}{|(n+1/2)h - x|} \int_{-y_n}^{y_n} |f((n+1/2)h + iy)| \, dy \to 0 \quad (2.31)$$

as $n \to \infty$. The limit on the right-hand side of (2.31) uses the condition (2.20). That the modulus of the second integral on the right-hand side of (2.28) goes to zero as $n \to \infty$ follows in exactly the same fashion. Equating (2.27) and (2.28), letting $n \to \infty$, and using (2.31) yields

$$\begin{aligned} \epsilon(x) &= f(x) - \sum_{k=-\infty}^{\infty} f(kh) \operatorname{sinc}\left(\frac{x-kh}{h}\right) \\ &= \lim_{n \to \infty} I_n(x) \\ &= S_h(x) \lim_{n \to \infty} \int_{-(n+1/2)h}^{(n+1/2)h} \left\{ \frac{F(x, t-iy_n)}{\sin\left(\frac{\pi}{h}(t-iy_n)\right)} \right. \quad (2.32) \\ &\quad \left. - \frac{F(x, t+iy_n)}{\sin\left(\frac{\pi}{h}(t+iy_n)\right)} \right\} dt \\ &= S_h(x) I(f, h)(x). \end{aligned}$$

To show (2.26), first consider $p = 1$. Note that

$$\begin{aligned} &|\sin(\pi(t \pm id)/h)| \\ &= |\sin(\pi t/h) \cosh(\pi d/h) \pm i \cos(\pi t/h) \sinh(\pi d/h)| \quad (2.33) \\ &\geq \sinh(\pi d/h) \end{aligned}$$

and $|t - x \pm id| \geq d$. Upon taking absolute values in (2.32),

$$\begin{aligned} |\epsilon(x)| &\leq \frac{1}{2\pi d \sinh(\pi d/h)} \int_{-\infty}^{\infty} \{|f(t + id^-)| + |f(t - id^-)|\} dt \\ &= \frac{N^1(f, D_S)}{2\pi d \sinh(\pi d/h)} \end{aligned}$$

if $f \in B^1(D_S)$. This implies (2.26) for $p = 1$.

In the case that $p = 2$, again take the absolute value of (2.32) to arrive at the inequality

$$|\epsilon(x)| \leq \frac{1}{2\pi \sinh(\pi d/h)} \int_{-\infty}^{\infty} \{|F(x, t-id^-)| + |F(x, t+id^-)|\} dt.$$

Since $f \in B^2(D_S)$, the Cauchy-Schwarz inequality (1.11) gives

$$\begin{aligned} &\left\{ \int_{-\infty}^{\infty} \left| \frac{f(t \pm id^-)}{t - x \pm id^-} \right| dt \right\} \\ &\leq \left\{ \int_{-\infty}^{\infty} \frac{dt}{|t - x \pm id^-|^2} \right\}^{1/2} \left\{ \int_{-\infty}^{\infty} |f(t \pm id^-)|^2 dt \right\}^{1/2} \\ &= \sqrt{\frac{\pi}{d}} \left\{ \int_{-\infty}^{\infty} |f(t \pm id^-)|^2 dt \right\}^{1/2}. \end{aligned}$$

This yields (2.26) for $p = 2$. From these the error in (2.26) satisfies

$$\begin{aligned}\|\epsilon\|_\infty &\equiv \sup_{x \in \mathbb{R}} |f(x) - C(f,h)(x)| \\ &\leq \frac{N^p(f,D_S)}{2(\pi d)^{1/p} \sinh(\pi d/h)} \\ &\leq \frac{2N^p(f,D_S)}{(\pi d)^{1/p}} e^{-\pi d/h}\end{aligned} \qquad (2.34)$$

where the last inequality is valid for all $h \leq 2\pi d/(\ell n(2))$. The latter is not a serious restriction since h (the mesh size) will always have the limiting value zero (see (2.41)). ■

The *truncated cardinal series* is denoted

$$C_{M,N}(f,h)(x) \equiv \sum_{k=-M}^{N} f(kh) \operatorname{sinc}\left(\frac{x - kh}{h}\right).$$

The previous examples had $M = N$, but this was due to the symmetry of the functions in the illustrations. The development is the same whether the sum is symmetric or not, and there are definite advantages, in the sense of computational efficiency, to having $M \neq N$. As indicated in Example 2.10, without further restrictions on the rate of growth of f on $\mathbb{R}$, the error in sinc approximation and the error in this truncated cardinal series may be unbalanced. The necessary restrictions on f, and the corresponding parameter selections that guarantee the exponential convergence of $C_{M,N}(f,h)$ to f, are summarized in the following theorem.

Theorem 2.14 *Assume $f \in B^p(D_S)$ ($p = 1$ or 2) and that there are positive constants α, β, and C so that*

$$|f(x)| \leq C \begin{cases} \exp(-\alpha|x|), & x \in (-\infty, 0) \\ \exp(-\beta|x|), & x \in [0, \infty). \end{cases} \qquad (2.35)$$

Make the selections

$$N = \left[\left|\frac{\alpha}{\beta} M + 1\right|\right] \qquad (2.36)$$

(where $[\![\cdot]\!]$ denotes the greatest integer function) and

$$h = (\pi d/(\alpha M))^{1/2} \leq (2\pi d)/(\ell n(2)) \qquad (2.37)$$

for the truncated cardinal expansion $C_{M,N}(f,h)$. Then

$$\|f - C_{M,N}(f,h)\|_\infty \leq K_1 M^{1/2} \exp(-(\pi d \alpha M)^{1/2}) \qquad (2.38)$$

where K_1 is a constant depending on f, p, and d.

Proof Consider the computation

$$\left| \sum_{k=-\infty}^{-M-1} f(kh) \, \text{sinc}\left(\frac{x-kh}{h}\right) \right|$$

$$\leq \sum_{k=M+1}^{\infty} |f(-kh)|$$

$$\leq C \sum_{k=M+1}^{\infty} e^{-\alpha kh} \qquad (2.39)$$

$$= C \left\{ \frac{e^{-\alpha h(M+1)}}{1-e^{-\alpha h}} \right\}$$

$$= C e^{-\alpha M h} \left\{ \frac{1}{e^{\alpha h}-1} \right\}$$

$$\leq \frac{C}{\alpha h} \exp(-\alpha M h).$$

In exactly the same fashion

$$\left| \sum_{k=N+1}^{\infty} f(kh) \, \text{sinc}\left(\frac{x-kh}{h}\right) \right| \leq \frac{C}{\beta h} \exp(-\beta N h). \qquad (2.40)$$

Using (2.34) with $h \leq (2\pi d)/(\ell n(2))$ and taking the absolute value of (2.22) yields

$$|f(x) - C_{M,N}(f,h)(x)|$$

$$\leq \sum_{k=M+1}^{\infty} |f(-kh)| + \sum_{k=N+1}^{\infty} |f(kh)| + \frac{2N^p(f,D_S)}{(\pi d)^{1/p}} e^{-\pi d/h}.$$

This leads, with (2.39) and (2.40), to the inequality

$$|f(x)-C_{M,N}(f,h)(x)| \leq \frac{C}{\alpha h}e^{-\alpha Mh}+\frac{C}{\beta h}e^{-\beta Nh}+\frac{2N^p(f,D_S)}{(\pi d)^{1/p}}e^{-\pi d/h}.$$

Making the selections in (2.36) and (2.37) allows one to see that

$$|f(x)-C_{M,N}(f,h)(x)|$$
$$\leq \left[\frac{C}{\alpha}+\frac{C}{\beta}+\frac{2N^p(f,D_S)}{(\pi d)^{1/p}}\sqrt{\left(\frac{\pi d}{\alpha M}\right)}\right]\sqrt{\left(\frac{\alpha}{\pi d}\right)}M^{1/2}e^{-(\pi d\alpha M)^{1/2}}$$
$$\leq \left[\frac{C}{\alpha}+\frac{C}{\beta}+\frac{2N^p(f,D_S)}{(\pi d)^{1/p}}\sqrt{\left(\frac{\pi d}{\alpha}\right)}\right]\sqrt{\left(\frac{\alpha}{\pi d}\right)}M^{1/2}e^{-(\pi d\alpha M)^{1/2}}$$
$$\equiv K_1 M^{1/2}e^{-(\pi d\alpha M)^{1/2}}.$$

∎

In the previous two theorems the inequality

$$h \leq \kappa\pi d/(\ell n(2)), \quad \kappa=1,2$$

was used to obtain

$$\frac{1}{\sinh(\kappa\pi d/h)} \leq 4\exp(-\kappa\pi d/h). \tag{2.41}$$

The form of the second inequality is conducive to balancing the contribution to the error that arises from the truncation of the infinite sum in interpolation or integration. The first inequality is almost always satisfied with the mesh selection

$$h = (\kappa\pi d/(\alpha M))^{1/2}$$

since, in this case, the requirement reduces to

$$\ell n(2) \leq (\alpha M)^{1/2}.$$

The selection N in (2.36) is for technical accuracy. If $\alpha M/\beta$ is not an integer, then the integer portion of $\alpha M/\beta$ is a bit small to balance the error contribution from the truncation of the upper sum. On the other hand, $\alpha M/\beta$ is a bit large if $\alpha M/\beta$ is an integer. Throughout the remainder of the text the selection in (2.36) will be used for technical accuracy but, for instance, in the various examples where $\alpha M/\beta$ is an integer this integer value will be used.

In many applications it is desirable to find approximations of $f(x)$ so that the k-th derivative of the approximation accurately approximates $\frac{d^k}{dx^k}f(x)$ (collocation for differential equations). To approximate the derivatives of f accurately, the representation in (2.22) is differentiated to arrive at the following theorem found in [5].

Theorem 2.15 *Let $f \in B^p(D_S)$, $(p = 1$ or $2)$ with $h > 0$. The interpolation error*

$$f(x) - C(f,h)(x) \equiv \epsilon(x) = S_h(x)I(f,h)(x)$$

has the n-th derivative

$$\frac{d^n \epsilon(x)}{dx^n} = \sum_{j=0}^{n} \binom{n}{j} \frac{d^j S_h(x)}{dx^j} \frac{d^{n-j} I(f,h)(x)}{dx^{n-j}}.$$

Thus for $p = 1$, if $h \leq \pi d$ then

$$\left\| \frac{d^n f}{dx^n} - \frac{d^n C(f,h)}{dx^n} \right\|_{\infty} \leq \frac{n!\, e N^1(f, D_S)}{2\pi d\, \sinh(\pi d/h)} \left(\frac{\pi}{h}\right)^n \qquad (2.42)$$

and for $p = 2$, if $h < \pi$, letting $\bar{d} = \max\{1, d^{-n}\}$, then

$$\left\| \frac{d^n f}{dx^n} - \frac{d^n C(f,h)}{dx^n} \right\|_{\infty}$$

(2.43)

$$\leq \frac{\bar{d} n!\, e^{1/2} N^2(f, D_S)}{2(\pi d)^{1/2} \sinh(\pi d/h)} \left(\frac{\pi}{h}\right)^n (1 - h^2/\pi^2)^{-1/4}.$$

Both (2.42) and (2.43) are of the order $\mathcal{O}(e^{-\pi d/h})$.

Proof The formula

$$\frac{d^n}{dx^n}(f(x)g(x)) = \sum_{j=0}^{n} \binom{n}{j} f^{(n-j)}(x) g^{(j)}(x)$$

applied to the product

$$\sin(\pi x/h) \cdot \frac{1}{(t - x \pm id)}$$

in (2.22) leads to the identity

$$f^{(n)}(x) - \sum_{k=-\infty}^{\infty} f(kh) \frac{d^n}{dx^n}\left[\operatorname{sinc}\left(\frac{x-kh}{h}\right)\right]$$

$$\equiv \frac{d^n}{dx^n} \epsilon(x)$$

$$= \frac{1}{2\pi i} \sum_{j=0}^{n} \binom{n}{j} j! \frac{d^{n-j}}{dx^{n-j}} (\sin(\pi x/h)) I_G(x)$$

where

$$I_G(x) = \int_{-\infty}^{\infty} \left\{ \frac{G(t-id^-)}{(t-x-id^-)^{j+1}} - \frac{G(t+id^-)}{(t-x+id^-)^{j+1}} \right\} dt$$

and

$$G(u \pm iv) = \frac{f(u \pm iv)}{\sin\left(\frac{\pi}{h}(u \pm iv)\right)}.$$

Note that

$$\left|\frac{d^{n-j}}{dx^{n-j}}(\sin(\pi x/h))\right| \leq \left(\frac{\pi}{h}\right)^{n-j}.$$

Then for $p=1$, using (2.33) and $|t-x \pm id|^{j+1} \geq d^{j+1}$ yields

$$|I_G(x)| \leq \frac{N^1(f,D_S)}{d^{j+1}\sinh(\pi d/h)}.$$

Thus

$$\left|\frac{d^n\epsilon(x)}{dx^n}\right| \leq \frac{1}{2\pi}\sum_{j=0}^{n}\binom{n}{j}\frac{j!}{d^{j+1}}\left(\frac{\pi}{h}\right)^{n-j}\frac{N^1(f,D_S)}{\sinh(\pi d/h)}$$

$$= \frac{n! N^1(f,D_S)}{2\pi d^{n+1}\sinh(\pi d/h)}\sum_{k=0}^{n}\left(\frac{1}{k!}\right)\left(\frac{\pi d}{h}\right)^k$$

$$\leq \frac{n!e N^1(f,D_S)}{2\pi d \sinh(\pi d/h)}\left(\frac{\pi}{h}\right)^n$$

since $\pi d/h \geq 1$.

For $p = 2$, the Cauchy–Schwarz inequality (1.11) yields

$$\int_{-\infty}^{\infty} \left| \frac{f(t \pm id^-)}{(t - x \pm id^-)^{j+1}} \right| dt$$

$$\leq \left\{ \int_{-\infty}^{\infty} \frac{dt}{|t - x \pm id^-|^{2j+2}} \right\}^{1/2} \left\{ \int_{-\infty}^{\infty} |f(t \pm id^-)|^2 dt \right\}^{1/2}$$

$$= \left[\frac{\pi \left(\frac{1}{2}\right)_j}{j! \, d^{2j+1}} \right]^{1/2} \left\{ \int_{-\infty}^{\infty} |f(t \pm id^-)|^2 dt \right\}^{1/2}$$

where repeated integration by parts has been applied and the Pochhammer symbol is defined by the ascending factorial

$$(a)_j = a(a+1)\cdots(a+j-1).$$

Hence

$$|I_G(x)| \leq \frac{\left[\pi \left(\frac{1}{2}\right)_j \right]^{1/2} N^2(f, D_S)}{(j!)^{1/2} d^{j+1/2} \sinh(\pi d/h)}$$

so that a calculation similar to the above yields

$$\left| \frac{d^n \epsilon(x)}{dx^n} \right| \leq \frac{1}{2\pi} \sum_{j=0}^{n} \binom{n}{j} (j!)^{1/2} \left(\frac{\pi}{h}\right)^{n-j} \frac{\left[\pi \left(\frac{1}{2}\right)_j \right]^{1/2} N^2(f, D_S)}{d^{j+1/2} \sinh(\pi d/h)}$$

$$\leq \frac{\bar{d} n! \, N^2(f, D_S)}{2(\pi d)^{1/2} \sinh(\pi d/h)} \left(\frac{\pi}{h}\right)^n$$

$$\times \sum_{j=0}^{n} \left\{ \left[\frac{\left(\frac{1}{2}\right)_j}{j!} \left(\frac{h}{\pi}\right)^{2j} \right]^{1/2} \left[\left(\frac{1}{(n-j)!}\right)^2 \right]^{1/2} \right\}$$

$$\leq \frac{\bar{d} n! \, N^2(f, D_S)}{2(\pi d)^{1/2} \sinh(\pi d/h)} \left(\frac{\pi}{h}\right)^n$$

$$\times \left[\sum_{j=0}^{n} \frac{\left(\frac{1}{2}\right)_j \left(\left(\frac{h}{\pi}\right)^2\right)^j}{j!} \right]^{1/2} \left[\sum_{k=0}^{n} \left(\frac{1}{k!}\right)^2 \right]^{1/2}$$

$$\leq \frac{\bar{d} n! \, e^{1/2} N^2(f, D_S)}{2(\pi d)^{1/2} \sinh(\pi d/h)} \left(\frac{\pi}{h}\right)^n \left(1 - \frac{h^2}{\pi^2}\right)^{-1/4}.$$

In the last inequality the binomial theorem was used in the form

$$\sum_{j=0}^{\infty} \frac{\left(\frac{1}{2}\right)_j \left(\left(\frac{h}{\pi}\right)^2\right)^j}{j!} = \sum_{j=0}^{\infty} \binom{-\frac{1}{2}}{j} \left(\frac{-h^2}{\pi^2}\right)^j$$
$$= \left(1 - \frac{h^2}{\pi^2}\right)^{-1/2}$$

which is valid for $h < \pi$.

∎

The following theorem provides the framework for truncating the infinite series approximation in Theorem 2.15. This result is also found in [5].

Theorem 2.16 *Assume that $f \in B^p(D_S)$ ($p = 1$ or 2) and that there are positive constants α, β, and C so that*

$$|f(x)| \le C \begin{cases} \exp(-\alpha|x|), & x \in (-\infty, 0) \\ \exp(-\beta|x|), & x \in [0, \infty). \end{cases}$$

Make the selections

$$N = \left[\left|\frac{\alpha}{\beta}M + 1\right|\right]$$

and

$$h = (\pi d/(\alpha M))^{1/2} \le \min\{\pi d, \pi/\sqrt{2}\}$$

for the n-th derivative of the truncated cardinal expansion, $\frac{d^n}{dx^n} C_{M,N}(f, h)(x)$. Then

$$\left\|\frac{d^n f}{dx^n} - \frac{d^n}{dx^n} C_{M,N}(f, h)\right\|_{\infty}$$

$$\le K_2 M^{(n+1)/2} \exp\left(-(\pi d\alpha M)^{1/2}\right)$$

(2.44)

where K_2 is a constant depending on f, n, p, and d.

Proof Differentiating the identity

$$\operatorname{sinc}\left(\frac{x-kh}{h}\right) = \frac{h}{2\pi}\int_{-\pi/h}^{\pi/h} e^{-ixt} e^{ikht} dt$$

and taking the absolute value of this result yields

$$\left|\frac{d^n}{dx^n} S(k,h)(x)\right| \leq \frac{1}{n+1}\left(\frac{\pi}{h}\right)^n.$$

Thus

$$\left|\frac{d^n f(x)}{dx^n} - \frac{d^n}{dx^n} C_{M,N}(f,h)(x)\right| \leq \left|\frac{d^n \epsilon(x)}{dx^n}\right|$$
$$+ \frac{C}{\alpha h(n+1)}\left(\frac{\pi}{h}\right)^n e^{-\alpha M h} + \frac{C}{\beta h(n+1)}\left(\frac{\pi}{h}\right)^n e^{-\beta N h}.$$

From Theorem 2.15 for $p = 1$ or 2,

$$\left|\frac{d^n \epsilon(x)}{dx^n}\right| \leq \frac{\bar{d}n!\, e^{1/p} N^p(f,D_S)}{2(\pi d)^{1/p} \sinh(\pi d/h)}\left(\frac{\pi}{h}\right)^n \left(1 - \frac{h^2}{\pi^2}\right)^{(1-p^2)/(3p^2)}$$
$$\leq \frac{4\bar{d}n!\, e^{1/p} N^p(f,D_S)}{(\pi d)^{1/p}}\left(\frac{\pi}{h}\right)^n e^{-\pi d/h}$$

where $h \leq \min\{\pi d, \pi/\sqrt{2}\}$ guarantees $h \leq 2\pi d/(\ln(2))$. Proceeding as in Theorem 2.14 yields the result with

$$K_2 \equiv \left[\left(\frac{1}{\alpha} + \frac{1}{\beta}\right)\frac{C}{n+1} \right.$$
$$\left. + \frac{4\bar{d}n!\, e^{1/p} N^p(f,D_S)}{(\pi d)^{1/p}}\left(\frac{\pi d}{\alpha}\right)^{1/2}\right]\left(\frac{\alpha}{d}\right)^{(n+1)/2} \pi^{(n-1)/2}.$$

∎

Example 2.17 Consider solving the boundary value problem

$$-u''(x) + q(x)u(x) = f(x), \quad -\infty < x < \infty \tag{2.45}$$

$$\lim_{x \to \pm\infty} u(x) = 0$$

via a collocation procedure with a sinc basis. That is, put

$$u_m(x) \equiv \sum_{k=-M}^{M} u_k \operatorname{sinc}\left(\frac{x - kh}{h}\right), \quad m = 2M + 1 \qquad (2.46)$$

where $\{u_k\}$ are unknown. Evaluating (2.45) at $x_j = jh$ yields

$$-u''(x_j) + q(x_j)u(x_j) = f(x_j).$$

From Theorem 2.16 (with $n = 2$ and $M = N$) one finds the relation

$$u''(jh) = \frac{1}{h^2} \sum_{k=-M}^{M} u(kh)\delta_{k,j}^{(2)} + \mathcal{O}\left(M^{3/2} \exp\left(-(\pi d \alpha M)^{1/2}\right)\right)$$

where Exercise 2.5 shows that

$$\begin{aligned}\delta_{k,j}^{(2)} &\equiv h^2 \frac{d^2}{dx^2}\left[\operatorname{sinc}\left(\frac{x - kh}{h}\right)\right]\bigg|_{x=jh} \\ &= \begin{cases} -\pi^2/3, & k = j \\ \frac{-2(-1)^{j-k}}{(j-k)^2}, & k \neq j. \end{cases}\end{aligned} \qquad (2.47)$$

Denoting $u(x_j)$ by u_j, one finds the linear system

$$\left[-\frac{1}{h^2} I^{(2)} + D(q)\right] \vec{u} = D(f)\vec{1} \qquad (2.48)$$

for the unknowns $\vec{u} \equiv (u_{-M}, \ldots, u_M)^T$. The kj-th component of $I^{(2)}$ is $\delta_{k,j}^{(2)}$ given in (2.47), $\vec{1} = (1, \ldots, 1)^T$, and the diagonal matrix

$$D(q) \equiv \begin{bmatrix} q(x_{-M}) & & \\ & \ddots & \\ & & q(x_M) \end{bmatrix}.$$

The convergence of the sinc-collocation method is discussed in Chapter 5. For a numerical experiment see Exercise 2.6.

■

It was seen in Example 2.11 that the trapezoidal rule of integration performed significantly better on the function $f_2(t) = \text{sech}(t)$ than it did on $f_1(t) = \exp(-|t|)$. The class of functions where the trapezoidal rule can be expected to give very accurate results is $B(D_S)$ of Definition 2.12. Before developing the error in the trapezoidal method of integration, a lemma is needed. The lemma also provides a nice application of contour integration.

Lemma 2.18 *If $\alpha > 0$ and $Im(z_0) \neq 0$, then*

$$\int_{-\infty}^{\infty} \frac{e^{i\alpha x}}{x - z_0} dx = \begin{cases} 2\pi i e^{i\alpha z_0}, & Im(z_0) > 0 \\ 0, & Im(z_0) < 0. \end{cases} \quad (2.49)$$

Proof Integrate

$$f(z) = \frac{e^{i\alpha z}}{z - z_0}$$

around the contour $\gamma = [-r, r] \cup \gamma_r$ ($r > |z_0|$), where $\gamma_r = \{z \in \mathbb{C} : z = re^{i\theta}, 0 < \theta < \pi\}$ (same contour as in Example 1.23). By the Residue Theorem 1.13,

$$2\pi i \begin{Bmatrix} e^{i\alpha z_0}, & Im(z_0) > 0 \\ 0, & Im(z_0) < 0 \end{Bmatrix} = \int_\gamma f(z) dz$$
$$\quad (2.50)$$
$$= \int_{-r}^{r} \frac{e^{i\alpha x}}{x - z_0} dx + \int_0^\pi \frac{\exp(i\alpha r e^{i\theta})}{re^{i\theta} - z_0} ire^{i\theta} d\theta.$$

Now $|re^{i\theta} - z_0| \geq r - |z_0|$ and $|\exp(i\alpha re^{i\theta})| = e^{-r\alpha \sin \theta}$, so the second integral on the right-hand side of (2.50) is bounded by

$$\left| \int_0^\pi \frac{\exp(i\alpha r e^{i\theta})}{re^{i\theta} - z_0} ire^{i\theta} d\theta \right| \leq \frac{r}{r - |z_0|} \int_0^\pi e^{-r\alpha \sin \theta} d\theta$$

$$= \frac{r}{r - |z_0|} \left\{ \int_0^{\pi/2} e^{-r\alpha \sin \theta} d\theta + \int_{\pi/2}^\pi e^{-r\alpha \sin \theta} d\theta \right\}$$

$$= \frac{2r}{r - |z_0|} \int_0^{\pi/2} e^{-r\alpha \sin \theta} d\theta$$

$$\leq \frac{2r}{r - |z_0|} \int_0^{\pi/2} e^{-r\alpha(2\theta/\pi)} d\theta$$

$$= \frac{\pi(1 - e^{-\alpha r})}{\alpha(r - |z_0|)} \to 0 \text{ as } r \to \infty.$$

Letting r tend to infinity in (2.50) yields (2.49).

∎

Corollary 2.19 *If h and d are positive, then*
$$\int_{-\infty}^{\infty} \frac{\sin(\pi x/h)}{t - x \pm id} dx = -\pi e^{i\pi \xi sgn(Im(\xi))/h}, \quad \xi = t \pm id$$
where
$$sgn(Im(\xi)) \equiv \begin{cases} +1, & Im(\xi) = d \\ -1, & Im(\xi) = -d. \end{cases}$$

Proof Write
$$\sin(\pi x/h) = \frac{e^{i\pi x/h} - e^{-i\pi x/h}}{2i}$$
and apply Lemma 2.18.

∎

The next theorem gives the error of the trapezoidal rule when applied to functions in $B(D_S)$.

Theorem 2.20 *If $f \in B(D_S)$ and $h > 0$, then*
$$\int_{-\infty}^{\infty} f(x)dx - h \sum_{k=-\infty}^{\infty} f(kh) \equiv \eta \qquad (2.51)$$
where
$$\eta = \int_{-\infty}^{\infty} \epsilon(x)dx$$
$$= \frac{e^{-\pi d/h}}{2i} \int_{-\infty}^{\infty} \left[\frac{f(t + id^-)e^{i\pi t/h}}{\sin(\pi(t + id^-)/h)} - \frac{f(t - id^-)e^{-i\pi t/h}}{\sin(\pi(t - id^-)/h)} \right] dt.$$
This integral representation is bounded by
$$|\eta| \leq \frac{N(f, D_S)}{2 \sinh(\pi d/h)} e^{-\pi d/h} \qquad (2.52)$$
so the quadrature error is of order $\mathcal{O}(e^{-2\pi d/h})$.

Proof Integrate both sides of (2.22) and use Corollary 2.19 to obtain η. The inequality in (2.52) follows from (2.33) and (2.21). ∎

The *truncated trapezoidal rule* is denoted

$$T_{M,N}(f,h) \equiv h \sum_{k=-M}^{N} f(kh).$$

The next theorem gives the quadrature error for functions in $B(D_S)$.

Theorem 2.21 *Assume $f \in B(D_S)$ and that there are positive constants α, β, and C so that*

$$|f(x)| \leq C \begin{cases} \exp(-\alpha|x|), & x \in (-\infty, 0) \\ \exp(-\beta|x|), & x \in [0, \infty). \end{cases}$$

Make the selections

$$N = \left[\left|\frac{\alpha}{\beta} M + 1\right|\right] \qquad (2.53)$$

and

$$h = (2\pi d/(\alpha M))^{1/2} \leq (2\pi d)/(\ell n(2)) \qquad (2.54)$$

for the truncated trapezoidal rule $T_{M,N}(f,h)$. Then

$$\left| \int_{-\infty}^{\infty} f(x)dx - h \sum_{k=-M}^{N} f(kh) \right| \leq K_3 \exp(-(2\pi d\alpha M)^{1/2}) \qquad (2.55)$$

where K_3 is a constant depending on f.

Proof Rewrite (2.51) in the form

$$\int_{-\infty}^{\infty} f(x)dx - h \sum_{k=-M}^{N} f(kh)$$

$$= h \sum_{k=N+1}^{\infty} f(kh) + h \sum_{k=M+1}^{\infty} f(-kh) + \eta.$$

Computations similar to the one in (2.39) show that

$$h \sum_{k=N+1}^{\infty} |f(kh)| \leq hC \sum_{k=N+1}^{\infty} e^{-\beta Nh}$$

$$\leq \frac{C}{\beta} \exp(-\beta Nh)$$

and

$$h \sum_{k=M+1}^{\infty} |f(-kh)| \leq \frac{C}{\alpha} \exp(-\alpha Mh).$$

Combining these with (2.52) (recalling (2.34)) yields

$$\left| \int_{-\infty}^{\infty} f(x)dx - h \sum_{k=-M}^{N} f(kh) \right|$$

$$\leq \frac{C}{\alpha} e^{-\alpha Mh} + \frac{C}{\beta} e^{-\beta Nh} + 2N(f, D_S) e^{-2\pi d/h}$$

$$\leq \left[\frac{C}{\alpha} + \frac{C}{\beta} + 2N(f, D_S) \right] \exp(-(2\pi d\alpha M)^{1/2})$$

$$\equiv K_3 \exp(-(2\pi d\alpha M)^{1/2}).$$

∎

Example 2.22 A Galerkin method using a sinc basis to solve the boundary value problem in (2.45)

$$Lu(x) \equiv -u''(x) + q(x)u(x) = f(x), \quad -\infty < x < \infty$$
(2.56)
$$\lim_{x \to \pm\infty} u(x) = 0$$

proceeds as follows. Assume an approximate solution of (2.56) in the form

$$u_m(x) \equiv \sum_{j=-M}^{M} u_j S(j, h)(x), \quad m = 2M + 1$$

where the $\{u_j\}$ are the undetermined coefficients in the approximate solution. In order to determine them, note that

$$Ru(x) \equiv Lu(x) - f(x) = 0$$

so that in the $L^2(\mathbb{R})$ inner product

$$\int_{-\infty}^{\infty} Ru(x)S(k,h)(x)dx = 0, \quad k = -M,\ldots,M \qquad (2.57)$$

where, recalling (2.9),

$$S(k,h)(x) = \text{sinc}\left(\frac{x - kh}{h}\right).$$

Written out, (2.57) takes the form

$$\int_{-\infty}^{\infty} -u''(x)S(k,h)(x)dx + \int_{-\infty}^{\infty} q(x)u(x)S(k,h)(x)dx$$
$$= \int_{-\infty}^{\infty} f(x)S(k,h)(x)dx.$$

Two integrations by parts show that

$$\int_{-\infty}^{\infty} -u''(x)S(k,h)(x)dx = -\int_{-\infty}^{\infty} u(x)S''(k,h)(x)dx$$

if $[S(k,h)(x)u'(x)]|_{-\infty}^{\infty} = 0$. Hence, (2.57) takes the form

$$\int_{-\infty}^{\infty} -u(x)S''(k,h)(x)dx + \int_{-\infty}^{\infty} q(x)u(x)S(k,h)(x)dx$$
$$\qquad\qquad\qquad\qquad\qquad\qquad\qquad\qquad\qquad (2.58)$$
$$= \int_{-\infty}^{\infty} f(x)S(k,h)(x)dx.$$

If the hypotheses of Theorem 2.21 are applicable to the various integrands in (2.58) with u replaced by u_m, the resulting linear system is identical to that in (2.48), namely,

$$\left[-\frac{1}{h^2} I^{(2)} + D(q)\right] \vec{u} = D(f)\vec{1}.$$

So in this setting the sinc-collocation and Sinc-Galerkin techniques are equivalent. This unusual situation allows one to employ the advantages of both of these commonly used techniques. A more general setting where Sinc-Galerkin and sinc-collocation are equivalent (or closely related) is the subject of Section 5.1.

Exercises

Exercise 2.1 The numerical approximation of the integral

$$\int_{-\infty}^{\infty} \exp(-x^2)dx = \sqrt{\pi}$$

can, with appropriate mesh selection, be very accurately computed. Show that

$$\left| \int_{-\infty}^{\infty} e^{-x^2} dx - h \sum_{k=-M}^{M} e^{-k^2 h^2} \right| \le K \exp(-\pi M) \qquad (2.59)$$

if $h = (\pi/M)^{1/2}$. Notice that $\sqrt{M}$ on the right-hand side of (2.55) has been replaced by M. This more rapid convergence is due to the behavior of the integrand in (2.59).

To see the bound in (2.59), replace f in (2.52) by $\exp(-(x^2))$ and note that $|\exp(-(x+id)^2)| = |\exp(-(x-id)^2)|$. Hence for this example η is bounded by

$$\begin{aligned}
|\eta| &\le \frac{e^{-\pi d/h}}{\sinh(\pi d/h)} \int_{-\infty}^{\infty} \left| e^{-(x+id)^2} \right| dx \\
&\le \frac{e^{-\pi d/h}}{\sinh(\pi d/h)} \int_{-\infty}^{\infty} e^{-x^2} e^{d^2} dx \\
&= \frac{\sqrt{\pi} e^{-\pi d/h} e^{d^2}}{\sinh(\pi d/h)} \\
&\le 4\sqrt{\pi} \exp\left(\frac{-2\pi d}{h} + d^2 \right)
\end{aligned}$$

where (2.41) has been used to obtain the last inequality. Show that the exponent in the error is minimized at $d = \pi/h$. Note that letting d increase with decreasing h is appropriate since the original integrand is an entire function [1]. Hence with this choice,

$$|\eta| \le 4\sqrt{\pi} \exp\left(\frac{-\pi^2}{h^2} \right).$$

For the truncation error, show that

$$h \sum_{k=M+1}^{\infty} |f(kh)| + h \sum_{k=M+1}^{\infty} |f(-kh)| = 2h \sum_{k=M+1}^{\infty} e^{-k^2 h^2}$$
$$\le 2e^{-M^2 h^2}.$$

Equating the exponents in the right-hand sides of these two error expressions gives $h = \sqrt{\pi/M}$ so that (2.59) follows. Calculate the approximation to $\sqrt{\pi}$ with the above selections for $M = 4, 8$, and 12. The latter will be accurate to 15 digits. Contrast this with Exercise 3.5.

Exercise 2.2 A specific case of Theorem 2.4 is provided by $g(t) = (t^2 + 1)^{-1}$. Use Example 1.23 and Parseval's Theorem 1.25 to show that

$$k(z) = g(z)\left[1 - e^{-\pi/h}\left(\cos\left(\frac{\pi z}{h}\right) - z\sin\left(\frac{\pi z}{h}\right)\right)\right].$$

Exercise 2.3 Show that

$$\frac{1}{x-a} - C\left(\frac{1}{x-a}, h\right)(x) = \frac{1}{(x-a)}\left[\frac{\sin(\pi x/h)}{\sin(\pi a/h)}\right] \quad (2.60)$$

by letting $f(x) = (x-a)^{-1}$ and using the technique in Theorem 2.13 (assume $a \neq nh$). You may assume that

$$\lim_{n\to\infty} \int_{\partial R_n} \frac{dz}{(z-a)(z-x)\sin(\pi z/h)} = 0$$

where R_n is the domain in Theorem 2.13. Conclude that the function $C(1/(z-a), h)(z)$ has a removable singularity at $z = a$, while the function $f(z) = (z-a)^{-1}$ has a pole of order one. For example, in (2.60) replace a by i and then by $-i$ and subtract these two results to establish (2.16).

Exercise 2.4 If $h > 0$, show that

$$\sum_{k=-\infty}^{\infty} \operatorname{sinc}\left(\frac{x-kh}{h}\right) = 1, \quad x \in \mathbb{R}.$$

There is a computational proof based on the known development [3, page 23]

$$\sum_{k=1}^{\infty} \frac{(-1)^k}{x^2 - k^2} = \frac{\pi}{2x}\left[\csc(\pi x) - \frac{1}{\pi x}\right].$$

One could also begin with (2.27) (take $f(x) \equiv 1$). Showing

$$\lim_{n\to\infty} \int_{\partial R_n} \frac{dz}{(z-x)\sin(\pi z/h)} = 0$$

is not entirely trivial.

Exercise 2.5 Establish (2.47). This is purely computational, but it helps to write

$$\text{sinc}\left(\frac{x-kh}{h}\right) = (-1)^k \frac{h}{\pi} \frac{\sin(\pi x/h)}{(x-kh)}.$$

Exercise 2.6 Build the system (2.48) for the boundary value problem

$$-u''(x) + u(x) = 2\,\text{sech}^3(x), \quad -\infty < x < \infty$$

$$\lim_{x \to \pm\infty} u(x) = 0.$$

That is, $q(x) \equiv 1$ and $f(x) = 2\,\text{sech}^3(x)$ in (2.45). Select $h = \pi/\sqrt{2M}$ and solve the system (2.48) for $M = 4, 8, 16,$ and 32 for the unknowns $\vec{u} = (u_{-N}, \ldots, u_0, \ldots, u_N)^T$ in

$$u_m(x) = \sum_{k=-M}^{M} u_k \,\text{sinc}\left(\frac{x-kh}{h}\right), \quad m = 2M+1.$$

After computing the $\{u_k\}$, determine

$$\|u(kh) - u_k\|_\infty \equiv \max_{-M \le k \le M} |u(kh) - u_k| \quad (2.61)$$

where the true solution is given by $u(x) = \text{sech}(x)$. Finally, divide the interval $I_M = \left[-\pi\sqrt{M}, \pi\sqrt{M}\right]$ into $5M$ equal parts by defining

$$y_j = -\pi\sqrt{M} + j\,\frac{2\pi\sqrt{M}}{5M}, \quad j = 0, 1, \ldots, 5M.$$

For corresponding M's compute

$$\max_{0 \le j \le 5M} |u(y_j) - u_m(y_j)|$$

and compare this with the quantity obtained in (2.61). In some sense, this is a numerical verification for using the uniform norm.

Exercise 2.7 Complete the development of the linear system in Example 2.22 by applying the trapezoidal quadrature rule (Theorem 2.21) to the three integrals in (2.58). The latter two of these integrals yield point evaluations and, therefore, diagonal matrices. For the first integral, it is handy to recall (2.47).

References

[1] N. Eggert and J. Lund, "The Trapezoidal Rule for Analytic Functions of Rapid Decrease," *J. Comput. Appl. Math.*, 27 (1989), pages 389–406.

[2] D. S. Gilliam, J. R. Lund, and C. F. Martin, "A Discrete Sampling Inversion Scheme for the Heat Equation," *Numer. Math.*, 54 (1989), pages 493–506.

[3] I. S. Gradshteyn and I. M. Ryzhik, *Table of Integrals, Series and Products*, Academic Press, Inc., New York, 1980.

[4] J. Lund, "Accuracy and Conditioning in the Inversion of the Heat Equation," in *Computation and Control*, Proc. Bozeman Conf. 1988, Progress in Systems and Control Theory, Vol. 1, Birkhäuser, Boston, 1989, pages 179–196.

[5] L. Lundin and F. Stenger, "Cardinal-Type Approximations of a Function and Its Derivatives," *SIAM J. Math. Anal.*, 10 (1979), pages 139–160.

[6] J. McNamee, F. Stenger, and E. L. Whitney, "Whittaker's Cardinal Function in Retrospect," *Math. Comp.*, 25 (1971), pages 141–154.

[7] F. Stenger, "Approximations via Whittaker's Cardinal Function," *J. Approx. Theory*, 17 (1976), pages 222–240.

[8] E. T. Whittaker, "On the Functions Which are Represented by the Expansions of the Interpolation-Theory," *Proc. Roy. Soc. Edinburgh*, 35 (1915), pages 181–194.

[9] J. M. Whittaker, *Interpolatory Function Theory*, Cambridge Tracts in Mathematics and Mathematical Physics, No. 33, Cambridge University Press, London, 1935.

Chapter 3

Numerical Methods on an Arc Γ

The numerical processes of interpolation, differentiation, and integration on the real line can, with the help of adroitly selected conformal maps, be adapted to handle these same processes on other subsets of the complex plane $\mathbb{C}$. In this chapter numerical techniques are derived for each of the aforementioned on an arc $\Gamma \in \mathbb{C}$. The derivation is the same whether one develops the techniques on an arbitrary arc Γ or restricts them to a particular setting. It is the case that all applications in this chapter as well as in the remainder of the text are on arcs Γ that are subsets of the real line $\mathbb{R}$. As in Chapter 2, interpolation theory on Γ is treated first. The bulk of the chapter is then contained in Section 3.2, which covers the important topic of numerical integration used for the inner product approximations of Chapter 4. As with any numerical process, there are scientific and artistic sides to the successful implementation of the process (in this case quadrature rules). Part of the bulk in Section 3.2 is created by a sequence of examples that permeate the section illustrating the rules' applicability and implementation. For the former, various rules of thumb are given to guide the user to a wise choice and for the latter, specific parameter selections to fine tune a rule are illustrated via examples. The final Section 3.3 covers those aspects of differentiation that play a role in Chapter 5.

3.1 Interpolation on Γ

Consider the approximation of the integral

$$I = \int_a^b F(u)du.$$

With the tools presently available, one mode of attack is to select a map

$$\psi : (-\infty, \infty) \to [a, b]$$

and consider

$$\begin{aligned} I &= \int_a^b F(u)du = \int_{-\infty}^{\infty} F(\psi(x))\psi'(x)dx \\ &= h \sum_{k=-\infty}^{\infty} F(\psi(kh))\psi'(kh) + \eta \end{aligned}$$

where the trapezoidal rule (2.51) has been used. If the hypotheses of Theorem 2.20 apply to the function $f = (F \circ \psi) \cdot \psi'$, then one expects very good results (exponential convergence). To see what sort of assumptions one may require on the function f, recall that the function must be an element of $B(D_S)$, which necessitates that f be analytic in D_S (Definition 2.12). If ψ is analytic in a domain containing $\mathbb{R}$ (D_S for example) and F is analytic in a domain D containing the interval (a, b), then f is analytic in D_S (recall the discussion at the end of Section 1.2). Thus one is led to consider maps ψ that are conformal. The following definition is fundamental to the development of sinc methods on arcs Γ in the complex plane. It is illustrated in Figure 3.1.

Definition 3.1 Let D be a domain in the $w = u + iv$ plane with boundary points $a \neq b$. Let $z = \phi(w)$ be a one-to-one conformal map of D onto the infinite strip

$$D_S \equiv \{z \in \mathbb{C} : z = x + iy, \ |y| < d\} \qquad (3.1)$$

where $\phi(a) = -\infty$ and $\phi(b) = +\infty$. Denote by $w = \psi(z)$ the inverse of the mapping ϕ and let

$$\Gamma \equiv \{w \in \mathbb{C} : w = \psi(x), \ x \in \mathbb{R}\} = \psi(\mathbb{R}). \qquad (3.2)$$

Let $B(D)$ denote the class of functions analytic in D which satisfy for some constant a with $0 \leq a < 1$,

$$\int_{\psi(x+L)} |F(w)dw| = \mathcal{O}(|x|^a), \quad x \to \pm\infty \tag{3.3}$$

where $L = \{iy : |y| < d\}$ and for γ a simple closed contour in D

$$N(F, D) \equiv \lim_{\gamma \to \partial D} \int_\gamma |F(w)dw| < \infty. \tag{3.4}$$

Further, for $h > 0$, define the nodes

$$w_k = \psi(kh), \quad k = 0, \pm 1 \pm 2, \ldots. \tag{3.5}$$

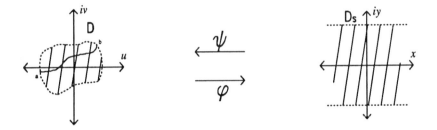

Figure 3.1 The conformally equivalent domains D and D_S.

The following interpolation theorem for F/ϕ', taken from [7], is a first step towards the sinc interpolation theorem for the function F.

Theorem 3.2 *Let $F \in B(D)$ and $h > 0$. Let ϕ be a one-to-one conformal map of the domain D onto D_S. If $\psi = \phi^{-1}$ and $w_k = \psi(kh)$, then for all $\xi \in \Gamma = \psi(\mathbb{R})$*

$$\epsilon(F)(\xi) \equiv \frac{F(\xi)}{\phi'(\xi)} - \sum_{k=-\infty}^{\infty} \frac{F(w_k)}{\phi'(w_k)} \operatorname{sinc}\left(\frac{\phi(\xi) - kh}{h}\right) \tag{3.6}$$

$$= \frac{\sin(\pi\phi(\xi)/h)}{2\pi i} \lim_{\gamma \to \partial D} \int_\gamma \frac{F(w)dw}{(\phi(w) - \phi(\xi))\sin(\pi\phi(w)/h)}.$$

Moreover,

$$\|\epsilon(F)\|_\infty \leq \frac{N(F, D)}{2\pi d \sinh(\pi d/h)}. \tag{3.7}$$

Proof Define the "conformal rectangles" $\psi(R_n)$, where

$$R_n = \left\{z \in \mathbb{C} : z = x + iy, |y| < y_n, -\left(n+\frac{1}{2}\right)h < x < \left(n+\frac{1}{2}\right)h\right\}$$

and $y_n = d - \frac{1}{n}$. Note that $R_n \subset D_S$. Then consider

$$\frac{\sin(\pi\phi(\xi)/h)}{2\pi i} \int_{\partial\psi(R_n)} \frac{F(w)dw}{(\phi(w)-\phi(\xi))\sin(\pi\phi(w)/h)}$$

$$= \frac{\sin(\pi x/h)}{2\pi i} \int_{\partial R_n} \frac{F(\psi(z))\psi'(z)dz}{(z-x)\sin(\pi z/h)}$$

$$= I_n(x)$$

where the change of variable $z = \phi(w)$ and $\phi(\xi) = x \in \mathbb{R}$ has been used. $I_n(x)$ is given in (2.27) where $f = (F \circ \psi) \cdot \psi'$. Application of the Residue Theorem 1.13 leads to

$$I_n(x) = F(\psi(x))\psi'(x) - \sum_{k=-n}^{n} F(\psi(kh))\psi'(kh)\,\text{sinc}\left(\frac{x-kh}{h}\right)$$

$$= \frac{F(\xi)}{\phi'(\xi)} - \sum_{k=-n}^{n} \frac{F(w_k)}{\phi'(w_k)}\,\text{sinc}\left(\frac{\phi(\xi)-kh}{h}\right).$$

The identities $\phi(\xi) = x$ if $\xi = \psi(x)$ and $\psi'(x) = [\phi'(\xi)]^{-1}$, as well as (3.5), give the second equality. The remainder of the development of (3.6) and (3.7) follows, as in the proof of Theorem 2.13.

∎

Corollary 3.3 *Let $F \in B(D)$ and $h > 0$. Let ϕ be a one-to-one conformal map of the domain D onto D_S. Let $\psi = \phi^{-1}$, $w_k = \psi(kh)$, and $\Gamma = \psi(\mathbb{R})$. Further assume that there are positive constants α, β, and C so that*

$$\left|\frac{F(\xi)}{\phi'(\xi)}\right| \leq C \begin{cases} \exp(-\alpha|\phi(\xi)|), & \xi \in \Gamma_a \\ \exp(-\beta|\phi(\xi)|), & \xi \in \Gamma_b \end{cases}$$

where

$$\Gamma_a \equiv \{\xi \in \Gamma : \phi(\xi) = x \in (-\infty, 0)\} \quad (3.8)$$

and
$$\Gamma_b \equiv \{\xi \in \Gamma : \phi(\xi) = x \in [0, \infty)\}. \tag{3.9}$$

If the selections
$$N = \left[\left|\frac{\alpha}{\beta} M + 1\right|\right] \tag{3.10}$$

and
$$h = \left(\frac{\pi d}{\alpha M}\right)^{1/2} \leq \frac{2\pi d}{\ln(2)} \tag{3.11}$$

are made, then for all $\xi \in \Gamma$ the error
$$\epsilon_{M,N}(F)(\xi) \equiv \frac{F(\xi)}{\phi'(\xi)} - \sum_{k=-M}^{N} \frac{F(w_k)}{\phi'(w_k)} \operatorname{sinc}\left(\frac{\phi(\xi) - kh}{h}\right)$$

is bounded by
$$\begin{aligned}
\|\epsilon_{M,N}(F)\|_\infty &\equiv \sup_{\xi \in \Gamma} |\epsilon_{M,N}(F)(\xi)| \\
&\leq K_4 M^{1/2} \exp\left(-(\pi d\alpha M)^{1/2}\right)
\end{aligned} \tag{3.12}$$

where K_4 is a constant depending on F, d, ϕ, and D.

Proof The proof is similar to that of Theorem 2.14. The restriction given by the inequality in (3.11) is again required because of (2.34), and this leads to
$$K_4 \equiv \left[\left(\frac{1}{\alpha} + \frac{1}{\beta}\right) C + \frac{2N(F, D)}{(\pi d\alpha)^{1/2}}\right] \left(\frac{\alpha}{\pi d}\right)^{1/2}.$$

∎

Notice that (3.6) is not in a "natural state" for the interpolation of $F(\xi)$. The formula (3.6) interpolates F/ϕ'. In this direction the following theorem is the applicable interpolation theorem.

Theorem 3.4 *Let $\phi' F \in B(D)$ and $h > 0$. Let ϕ be a one-to-one conformal map of the domain D onto D_S. Let $\psi = \phi^{-1}$, $w_k = \psi(kh)$, and $\Gamma = \psi(\mathbb{R})$. Then for all $\xi \in \Gamma$,*
$$\epsilon(\phi' F)(\xi) \equiv F(\xi) - \sum_{k=-\infty}^{\infty} F(w_k) \operatorname{sinc}\left(\frac{\phi(\xi) - kh}{h}\right)$$

is bounded by

$$\|\epsilon(\phi'F)\|_\infty \leq \frac{N(\phi'F, D)}{2\pi d \sinh(\pi d/h)}. \qquad (3.13)$$

Further assume that there are positive constants α, β, and C so that

$$|F(\xi)| \leq C \begin{cases} \exp(-\alpha|\phi(\xi)|), & \xi \in \Gamma_a \\ \exp(-\beta|\phi(\xi)|), & \xi \in \Gamma_b \end{cases} \qquad (3.14)$$

where Γ_q ($q = a, b$) are defined in (3.8) and (3.9). If the selections

$$N = \left[\left|\frac{\alpha}{\beta} M + 1\right|\right]$$

and

$$h = \left(\frac{\pi d}{\alpha M}\right)^{1/2} \leq \frac{2\pi d}{\ln(2)}$$

are made, then for all $\xi \in \Gamma$,

$$\epsilon_{M,N}(\phi'F)(\xi) \equiv F(\xi) - C_{M,N}(F, h, \phi)(\xi)$$

where

$$C_{M,N}(F, h, \phi)(\xi) \equiv \sum_{k=-M}^{N} F(w_k) \operatorname{sinc}\left(\frac{\phi(\xi) - kh}{h}\right) \qquad (3.15)$$

is bounded by

$$\|\epsilon_{M,N}(\phi'F)\|_\infty \leq K_5 M^{1/2} \exp\left(-(\pi d\alpha M)^{1/2}\right) \qquad (3.16)$$

and K_5 is a constant depending on F, d, ϕ, and D.

Proof Apply Theorem 3.2 and Corollary 3.3 to $\phi'F$. Then

$$K_5 \equiv \left[\left(\frac{1}{\alpha} + \frac{1}{\beta}\right) C + \frac{2N(\phi'F, D)}{(\pi d\alpha)^{1/2}}\right] \left(\frac{\alpha}{\pi d}\right)^{1/2}.$$

∎

In order to obtain a flavor of the previous result, consider the conformal equivalence of the domains

$$D_E \equiv \left\{ w \in \mathbb{C} : \left|\arg\left(\frac{w}{1-w}\right)\right| < d \leq \pi/2 \right\} \qquad (3.17)$$

and D_S defined by

$$z = \phi(w) = \ell n\left(\frac{w}{1-w}\right), \quad w = \psi(z) = \frac{e^z}{e^z + 1}, \qquad (3.18)$$

(see Exercise 1.1 and Example 1.16). Figure 3.2 displays the domain and range of the map ϕ via the composition of the conformal maps

$$\eta = \phi_1(w) = \frac{w}{1-w}$$

and

$$z = \phi_2(\eta) = \ell n(\eta).$$

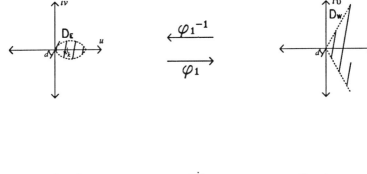

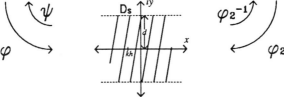

Figure 3.2 The domains D_E, D_W, and D_S.

Let $d = \pi/2$, in which case $D_E = \{w \in \mathbb{C} : |w - 1/2| < 1/2\}$. In order that $\phi' F$ belong to $B(D_E)$, the requirements (3.3) and (3.4) must be fulfilled, in addition to those of analyticity. To gain a feeling for these conditions in terms of the domain D_E, define

$$w = u + iv = \psi(x + iy) = \frac{e^x e^{iy}}{e^x e^{iy} + 1}$$

$$= \frac{e^{2x} + e^x \cos y}{A} + i \frac{e^x \sin y}{A}$$

(3.19)

where

$$A = e^{2x} + 2e^x \cos y + 1.$$

In Figure 3.3 the image of the two horizontal strips in D_S is of the upper and lower arcs of the circle $(u - 1/2)^2 + v^2 = 1/4$ (a short computation using (3.19) and $y = \pm \pi/2$). The image of the two vertical segments of the strip are the two "baseball seams" in D_E given by

$$\left(u - \frac{e^{2\bar{x}}}{B} \right)^2 + v^2 = \left(\frac{e^{\bar{x}}}{B} \right)^2$$

for $z = \bar{x} + iy$ and

$$\left(u + \frac{1}{B} \right)^2 + v^2 = \left(\frac{e^{\bar{x}}}{B} \right)^2$$

for $z = -\bar{x} + iy$, where

$$B = e^{2\bar{x}} - 1.$$

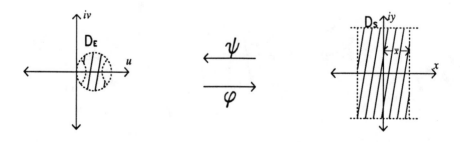

Figure 3.3 The domains D_E and D_S ($d = \pi/2$ and $|x| < \bar{x}$).

If these seams are denoted by B_S, then (3.3) reads for $\phi' F$,

$$\int_{B_S} |\phi'(w)F(w)dw| = \int_{B_S} \left|\frac{F(w)}{w(1-w)} dw\right|$$
$$= \mathcal{O}(|u|^a), \quad u \to 0^+, \; 1^-, \; 0 \leq a < 1.$$

The condition (3.4) for $\phi' F$ becomes

$$N(\phi' F, D_E) = \lim_{\gamma \to \partial D_B} \int_\gamma |\phi'(w)F(w)dw|$$
$$= \lim_{r \to 1/2^-} \int_0^{2\pi} r \left|\frac{F(1/2 + re^{i\theta})}{1/4 - r^2 e^{2i\theta}}\right| d\theta < \infty.$$

Condition (3.14) concerning the asymptotic behavior of F at $u = 0$ and $u = 1$ reads

$$|F(u)| \leq C \begin{cases} \left(\frac{u}{1-u}\right)^\alpha, & u \in (0, 1/2) \\ \left(\frac{u}{1-u}\right)^{-\beta}, & u \in [1/2, 1). \end{cases}$$

The general Theorem 3.4 takes the following form when specialized for the map $z = \ell n\left(\frac{w}{1-w}\right)$.

Theorem 3.5 Let $z = \phi(w)$ with $\phi(w) = \ell n\left(\frac{w}{1-w}\right)$. Let $\phi' F$ be analytic in D_E ($d = \pi/2$),

$$\int_{B_S} \left|\frac{F(w)}{w(1-w)} dw\right| = \mathcal{O}(|u|^a), \quad u \to 0^+, \; 1^- \; 0 \leq a < 1$$

and

$$N(\phi' F, D_E) = \lim_{r \to 1/2^-} \int_0^{2\pi} r \left|\frac{F(1/2 + re^{i\theta})}{1/4 - r^2 e^{2i\theta}}\right| d\theta < \infty.$$

Further assume that there are positive constants α, β, and C so that

$$|F(u)| \leq C \begin{cases} u^\alpha, & u \in (0, 1/2) \\ (1-u)^\beta, & u \in [1/2, 1) \end{cases}$$

If the selections

$$N = \left[\left|\frac{\alpha}{\beta} M + 1\right|\right], \quad h = \frac{\pi}{\sqrt{2\alpha M}} \leq \frac{\pi^2}{\ell n(2)}$$

are made, then

$$\left| F(u) - \sum_{k=-M}^{N} F(u_k) \operatorname{sinc} \left(\frac{\ell n \left(\frac{u}{1-u} \right) - kh}{h} \right) \right|$$
$$= \mathcal{O}\left(\sqrt{M} \exp(-\pi \sqrt{\alpha M/2}\,) \right)$$

where

$$u_k \equiv \psi(kh) = \frac{e^{kh}}{e^{kh} + 1}.$$

■

Three of the members of the sinc basis composed with the map $\phi(u) = \ell n(u/(1-u))$ are shown in Figure 3.4.

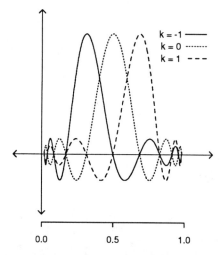

Figure 3.4 The sinc basis $(S(k,h) \circ \ell n(u/(1-u))$ on the interval $(0,1)$ for $k = -1, 0, 1, h = \pi/4)$.

Example 3.6 The function

$$F(u) = u^{1/3}(1-u)^{1/2} \qquad (3.20)$$

satisfies the hypotheses of Theorem 3.5 with $\alpha = 1/3$ and $\beta = 1/2$. Selecting $d = \pi/2$ so that $h = \pi\sqrt{3/(2M)}$ and $[|N = (2/3)M + 1|]$, the error in approximating F by the sinc expansion

$$C_{M,N}(F,h,\phi)(u) = \sum_{k=-M}^{N} F(u_k) \operatorname{sinc}\left(\frac{\phi(u) - kh}{h}\right)$$

yields an order estimate for the convergence rate given by

$$CR = \sqrt{M} \exp(-\pi\sqrt{M/6}).$$

There are a number of ways to represent the error of approximation of F by $C_{M,N}(F,h,\phi)$. A method that, for small M and N, numerically represents the order statement in Theorem 3.5 calculates the maximum absolute error on the uniform gridpoints

$$U = \{z_0, \ldots, z_{100}\}$$
$$z_j = jh_u, \quad h_u = .01, \quad k = 0, 1, \ldots, 100$$

with

$$\|E_U(h_u)\| = \max_{0 \leq j \leq 100} |F(z_j) - C_{M,N}(F,h,\phi)(z_j)|$$

and then approximates the order constant by the numerical order constant

$$NOC = \frac{\|E_U(h_u)\|}{CR}.$$

For a sequence of values of M, Table 3.1 lists the parameters N and h of the approximate solution in the first three columns, the numerical uniform error $\|E_U(h_u)\|$ in the fourth column and the numerical order constant in the final column. The values of NOC decrease as M increases due to the coarseness of the mesh. The uniform error is quite a bit smaller than the convergence rate CR since the former is measured with respect to the interval $[.01, .99]$ and there are, for example, sinc nodes $w_{-k} \in (0, .01)$ for $k \geq 4$. The same is true in the interval $(.99, 1)$. That is, the uniform error does not accurately reflect the error in the approximation at the endpoints of the interval where, due to the singularities of F, the error is likely to be the worst. For example, in the third line of Table 3.1, the leftmost sinc gridpoint is $w_{-8} = \exp(-\sqrt{8}h)/(1 - \exp(-\sqrt{8}h)) \simeq .0001$ and $w_{-4} = \exp(-2h)/(1 - \exp(-2h)) \simeq .01$. Hence, there are four gridpoints in a neighborhood of zero that are left out when calculating the error $\|E_U(h_u)\|$. Not only in this example, but in many examples that have been computed, the convergence rate CR is an accurate approximation of the error. Example 3.18 further highlights the convergence rate representing the error in the approximation.

M	N	$h = \pi\sqrt{3/(2M)}$	$\|E_U(h_u)\|$	NOC
2	2	2.721	$.191 \times 10^{-1}$	.083
4	3	1.924	$.739 \times 10^{-2}$	.048
8	6	1.360	$.375 \times 10^{-3}$	.005
16	11	0.962	$.128 \times 10^{-3}$	.005

Table 3.1 Error in sinc interpolation of the function (3.20).

∎

The result of Theorem 3.5 is in sharp contrast to the error in polynomial approximation of functions with algebraic singularities [5, pages 84–86]. Consider the function

$$F(u) = u^\alpha (1-u)^\beta$$

with positive nonintegral α and/or β. This function satisfies all of the hypotheses of Theorem 3.5. However, if this function with $\beta = \alpha \in (0,1)$ is approximated by polynomials of degree less than or equal to n (denoted Π_n), then there is a constant C such that

$$\inf_{q \in \Pi_n} \sup_{u \in [0,1]} |[u(1-u)]^\alpha - q(u)| \leq \frac{C}{n^\alpha}.$$

3.2 Quadrature on Γ

Just as the error in the trapezoidal method of integration on the entire real line was developed by integrating the interpolation error, the same methodology applies to obtain the error of integration over an arc $\Gamma = \psi(\mathbb{R})$. The development in this section follows the work in [6].

Theorem 3.7 *Let $F \in B(D)$ and $h > 0$. Let ϕ be a one-to-one conformal map of the domain D onto D_S. Let $\psi = \phi^{-1}$, $w_k = \psi(kh)$, and $\Gamma = \psi(\mathbb{R})$. Then*

$$\eta(F) \equiv \int_\Gamma F(\xi)d\xi - h \sum_{k=-\infty}^{\infty} \frac{F(w_k)}{\phi'(w_k)}$$
(3.21)
$$= \frac{i}{2} \lim_{\gamma \to \partial D} \int_\gamma \frac{F(w)\kappa(\phi,h)(w)}{\sin(\pi\phi(w)/h)} dw$$

where

$$\kappa(\phi,h)(w) \equiv \exp\left[\frac{i\pi\phi(w)}{h} sgn(Im(\phi(w)))\right].$$
(3.22)

Moreover

$$|\eta(F)| \leq \frac{N(F,D)}{2\sinh(\pi d/h)} e^{-\pi d/h}.$$
(3.23)

Proof Multiplying (3.6) by $\phi'(\xi)$ and integrating over Γ leads to the quantity defined by $\eta(F)$ in (3.21) above. For the error term in this expression, set $x = \phi(\xi)$ in the integral over Γ and, using Corollary 2.19, it follows that

$$\int_\Gamma \frac{\sin(\pi\phi(\xi)/h)\phi'(\xi)}{\phi(w)-\phi(\xi)} d\xi = \int_{-\infty}^{\infty} \frac{\sin(\pi x/h)}{-x + Re(\phi(w)) + i\, Im(\phi(w))} dx$$
$$= -\pi \exp\left[\frac{i\pi\phi(w)}{h} sgn(Im(\phi(w)))\right].$$

This gives the equality (3.21). The inequality in (3.23) follows from (3.21) using (2.33) and

$$|\kappa(\phi,h)(w)|\Big|_{w \in \partial D} = \left|\exp\left[\frac{i\pi\phi(w)}{h} sgn(Im(\phi(w)))\right]\right|\Big|_{w \in \partial D}$$
(3.24)
$$= e^{-\pi d/h}.$$

This equality follows from the fact that if $w \in \partial D$, then $\phi(w) = x \pm id$. ∎

Notice that the rule in (3.21) is of trapezoidal type. That is, it can be obtained from the trapezoidal rule by changing variables (recall (3.1) and $\psi = \phi^{-1}$). Some authors refer to such rules as mapped trapezoidal rules. The important point with regard to Theorem 3.7 is that the exponential convergence rate of the trapezoidal rule (2.51) is preserved under mappings that are conformal.

Theorem 3.8 *Let $F \in B(D)$ and $h > 0$. Let ϕ be a one-to-one conformal map of the domain D onto D_S. Let $\psi = \phi^{-1}$, $w_k = \psi(kh)$, and $\Gamma = \psi(\mathbb{R})$. Further assume that there are positive constants α, β, and C so that*

$$\left|\frac{F(\xi)}{\phi'(\xi)}\right| \leq C \begin{cases} \exp(-\alpha|\phi(\xi)|), & \xi \in \Gamma_a \\ \exp(-\beta|\phi(\xi)|), & \xi \in \Gamma_b \end{cases} \quad (3.25)$$

where Γ_q ($q = a, b$) are given in (3.8) and (3.9), respectively. Make the selections

$$N = \left[\left|\frac{\alpha}{\beta} M + 1\right|\right]$$

and

$$h = \left(\frac{2\pi d}{\alpha M}\right)^{1/2} \leq \frac{2\pi d}{\ell n 2}$$

for the truncated (mapped) quadrature rule

$$T_{M,N}(F,h,\phi) \equiv h \sum_{k=-M}^{N} \frac{F(w_k)}{\phi'(w_k)}.$$

Then

$$\eta_{M,N}(F) \equiv \int_\Gamma F(\xi)d\xi - T_{M,N}(F,h,\phi)$$

is bounded by

$$|\eta_{M,N}(F)| \leq K_6 \exp\left(-(2\pi d\alpha M)^{1/2}\right) \quad (3.26)$$

where K_6 is a constant depending only on F, d, ϕ, and D.

Proof Rewrite (3.21) as

$$\int_\Gamma F(\xi)d\xi - T_{M,N}(F,h,\phi)$$

$$(3.27)$$

$$= h \sum_{k=N+1}^{\infty} \frac{F(w_k)}{\phi'(w_k)} + h \sum_{k=M+1}^{\infty} \frac{F(w_{-k})}{\phi'(w_{-k})} + \eta(F).$$

One bounds the modulus of the truncated terms in (3.27) exactly as in Theorem 2.21 using (3.25). Hence, using (2.41) leads to

$$\left| \int_\Gamma F(\xi)d\xi - h \sum_{k=-M}^{N} \frac{F(w_k)}{\phi'(w_k)} \right|$$
$$\leq \frac{C}{\alpha} e^{-\alpha M h} + \frac{C}{\beta} e^{-\beta N h} + 2N(F,D)e^{-2\pi d/h}.$$

The selections for h and N give (3.26) where

$$K_6 \equiv \left(\frac{C}{\alpha} + \frac{C}{\beta} + 2N(F,D) \right).$$

∎

As was previously done for the $\ell n(\frac{w}{1-w})$ map, it is convenient to have the hypotheses of Theorem 3.8 sorted out for particular domains D and maps ϕ. For the domain D_E in (3.17) the development for quadratures is very similar to the development for interpolation. The details are given in Exercise 3.1.

Recalling Example 1.16, the transformation

$$z = \phi(w) = \ell n(w), \quad w = e^z$$

conformally maps

$$D_W \equiv \{w \in \mathbb{C} : w = re^{i\theta}, \ |\theta| < d \leq \pi/2\}$$

onto the infinite strip

$$D_S = \{z \in \mathbb{C} : z = x + iy, \ |y| < d \leq \pi/2\}.$$

The transcription of the conditions (3.3) and (3.4) for this domain (see Figure 3.5) is quite a bit simpler than it was for the mapping $z = \ell n(\frac{w}{1-w})$. Indeed, the image of the vertical segments $z = \pm \bar{x} + iy$ are the sectors of circles in the w-plane since

$$\psi(\pm\bar{x} + iy) = e^{\pm\bar{x}} e^{iy}.$$

Thus (3.3) becomes

$$\int_{\psi(x+L)} |F(w)dw| = \int_{-d}^{d} |F(re^{i\theta})| r \, d\theta$$
$$= \mathcal{O}(|\ell n(r)|^a), \quad r \to 0^+, \infty, \ 0 \leq a < 1.$$

To see if (3.4) holds, check the integral

$$\lim_{\gamma \to \partial D_W} \int_\gamma |F(w)dw| = \lim_{\substack{r \to 0^+ \\ R \to \infty}} \int_r^R |F(\rho e^{id})| d\rho < \infty.$$

A short computation shows that (3.25) is equivalent to

$$|F(u)| \leq C \begin{cases} u^{\alpha-1}, & u \in (0,1) \\ u^{-\beta-1}, & u \in [1,\infty). \end{cases}$$

The nodes of the quadrature formula u_k are given by

$$u_k = \psi(kh) = e^{kh}$$

and the weights $1/\phi'(u_k)$ are given by

$$\frac{1}{\phi'(u_k)} = e^{kh}.$$

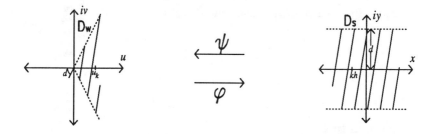

Figure 3.5 The conformally equivalent domains D_W and D_S.

The mapping $z = \phi(w) = \ln(w)$ plays a prevalent role in the remainder of this text. It played an unobtrusive role in Example 2.11. The details of Theorem 3.8 take the following form for this transformation.

Theorem 3.9 Let $z = \phi(w) = \ln(w)$. Let F be analytic in D_W where the angle of the wedge opening is d ($0 < d \leq \pi/2$),

$$\int_{-d}^{d} |F(re^{i\theta})| r d\theta = \mathcal{O}(|\ln r|^a), \quad r \to 0^+, \infty, \quad 0 \leq a < 1, \quad (3.28)$$

and
$$\lim_{\substack{r \to 0+ \\ R \to \infty}} \int_r^R |F(\rho e^{id})| d\rho < \infty. \tag{3.29}$$

Further assume that there are positive constants α, β, and C so that

$$|F(u)| \le C \begin{cases} u^{\alpha-1}, & u \in (0,1) \\ u^{-\beta-1}, & u \in [1,\infty). \end{cases} \tag{3.30}$$

If the selections

$$N = \left[\left|\frac{\alpha}{\beta} M + 1\right|\right] \tag{3.31}$$

and

$$h = \left(\frac{2\pi d}{\alpha M}\right)^{1/2} \le \frac{2\pi d}{\ell n 2} \tag{3.32}$$

are made, then

$$\left|\int_0^\infty F(u) du - h \sum_{k=-M}^N e^{kh} F(e^{kh})\right| = \mathcal{O}\left(\exp(-(2\pi d\alpha M)^{1/2})\right). \tag{3.33}$$

∎

Three members of the translated sinc basis composed with the conformal map $\phi(w) = \ell n(w)$ are shown in Figure 3.6.

As an illustration of the integrals in (3.28) and (3.29), consider the approximation of

$$I = \int_{-\infty}^\infty \frac{du}{u^2+1} = \pi$$

when using the trapezoidal rule (2.51). One cannot expect exponential convergence of this rule because of the rational behavior of the integrand at infinity. This has to do with the truncation error, which may be bounded by a calculation that is almost identical to the estimate worked out in (2.18). However, upon writing the integral as

$$I = 2\int_0^\infty \frac{du}{u^2+1},$$

then (3.30) is satisfied with $\alpha = \beta = 1$. The quantity d can be taken as $\pi/2 - \varepsilon$ for any positive ε, since the integrand has singularities at

$w = \pm i$. The exact computation of the integrals (3.29) and (3.30) is however, for arbitrary ε, quite difficult (one is led to elliptic integrals). The goal here is to remove a bit of the mystique from these integrals. A flavor for bounding these integrals is obtained if one selects $d = \pi/4$. For example, (3.29) becomes

$$\begin{aligned}
\int_{-d}^{d} \frac{r d\phi}{|r^2 e^{2i\phi} + 1|} &= 2r \int_{0}^{\pi/4} \frac{d\phi}{|r^2 e^{2i\phi} + 1|} \\
&= 2r \int_{0}^{\pi/4} \frac{d\phi}{[r^4 + 2r^2 \cos 2\phi + 1]^{1/2}} \\
&\leq \frac{2r}{(r^4 + 1)^{1/2}} (\pi/4) \to 0 \text{ as } r \to 0^+, \infty.
\end{aligned}$$

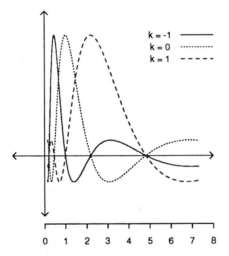

Figure 3.6 The sinc base $(S(k, h) \circ \ell n(u))$ on the interval $(0, \infty)$ for $k = -1, 0, 1, h = \pi/4$.

The integral in (3.29) takes the form

$$\begin{aligned}
\int_{r}^{R} |F(\rho e^{id})| d\rho &= \int_{r}^{R} \frac{d\rho}{|\rho^2 e^{2id} + 1|} \\
&= \int_{r}^{R} \frac{d\rho}{[\rho^4 + 2\rho^2 \cos 2d + 1]^{1/2}} \\
&\leq \int_{r}^{R} \frac{d\rho}{[\rho^4 + 1]^{1/2}}
\end{aligned}$$

$$\leq \int_0^1 \frac{d\rho}{[\rho^4+1]^{1/2}} + \int_1^\infty \frac{d\rho}{[\rho^4+1]^{1/2}}$$
$$\leq 1 + \int_1^\infty \frac{d\rho}{\rho^2} = 2.$$

In general one does not do the above calculation prior to implementing the rule (3.33). Rather, if the integrand F is analytic in a sector of the right half plane (angle of opening d) and integrable on rays in this sector, then this angle d defines, via (3.32), a suitable mesh size. The exponents α and β in (3.30) refer only to the behavior of F on the half line. In the present case the predicted asymptotic convergence rate is, by (3.33),

$$CR = \exp(-\pi\sqrt{M})$$

and the integral is approximated by the sum

$$2\int_0^\infty \frac{du}{u^2+1} = 2h \sum_{k=-M}^{M} \frac{e^{kh}}{e^{2kh}+1} = h \sum_{k=-M}^{M} \frac{1}{\cosh(kh)}.$$

This is the same sum as was used in Example 2.11 so that the errors listed in Table 2.1 are the same for this approximation.

The development of the formula (3.33) did not depend on the boundedness of F or its higher derivatives at $u = 0$. Indeed, for functions F with integrable singularities the convergence of the method can be more rapid than for an integrand that has no singularity. As Exercise 3.3 points out this rate is governed by α in (3.30).

Example 3.10 A more interesting integral than the example above is provided by

$$\int_0^\infty \left(\frac{1}{u+1} - e^{-u}\right) \frac{du}{u} = \lim_{n\to\infty} \left[\sum_{j=1}^n \frac{1}{j} - \ell n(n)\right] \equiv \gamma \qquad (3.34)$$

where $\gamma = .57721566490153\ldots$ is the Euler–Mascheroni constant. The integrand is analytic in the right half plane, so $d = \pi/2$ combined with (3.32) gives $h = \pi/\sqrt{M}$. Near the origin, the integrand is $\mathcal{O}(u)$, so by (3.30) the largest possible α is two. It is not necessary to obtain this optimal value (largest possible) of α for accurate computations. Noting that the integrand in (3.34) is bounded, then (3.30) is satisfied with $\alpha = 1$. This is used in this example, and the results obtained from the optimal selection of α are requested in Exercise 3.2. Although a portion of the integrand decreases exponentially at infinity, the dominant behavior is $\mathcal{O}(1/u^2)$. It may at first seem more efficient to split the integral into a sum of two integrals but neither of the resulting integrals converge. Hence, by (3.30) the selection $\beta = 1$ is made and $N = M$. In Table 3.2, each of these quantities is displayed along with the absolute error, the predicted asymptotic convergence rate, and the numerical order constant.

$M = N$	$h = \pi/\sqrt{M}$	AE	CR	NOC
4	1.571	$.423 \times 10^{-2}$	$.187 \times 10^{-2}$	2.27
8	1.111	$.210 \times 10^{-3}$	$.138 \times 10^{-3}$	1.52
16	0.785	$.238 \times 10^{-5}$	$.349 \times 10^{-5}$	0.68
32	0.555	$.123 \times 10^{-7}$	$.191 \times 10^{-7}$	0.64

Table 3.2 Sinc quadrature (3.33) applied to (3.34).

With the above selections of M, N, and h the predicted asymptotic convergence rate (from (3.33)) and the absolute error reported are

$$CR = \exp(-\pi\sqrt{M})$$

and

$$AE = \left| \int_0^\infty \left(\frac{1}{u+1} - e^{-u} \right) \frac{du}{u} - h \sum_{k=-M}^{M} e^{kh} F(e^{kh}) \right|,$$

respectively. The estimate of the numerical order constant is defined by $NOC = AE/CR$.

The previous example points to a situation in which one may improve the performance of the $\phi(w) = \ell n(w)$ quadrature rule with respect to the number of function evaluations. Specifically, assume the integrand exhibits exponential decrease at infinity

$$|F(u)| \leq C \begin{cases} u^{\alpha-1}, & u \in (0,1) \\ e^{-\beta u}, & u \in [1,\infty). \end{cases} \qquad (3.35)$$

The number of integration nodes is significantly reduced by selecting

$$N_e = \left[\left|\frac{1}{h} \ell n\left(\frac{\alpha}{\beta} Mh\right) + 1\right|\right]. \qquad (3.36)$$

That the exponential convergence of the approximating sum is maintained with this alternative selection of N_e (h the same as in (3.32)) is verified by the calculation

$$\begin{aligned}
h \sum_{k=N_e+1}^{\infty} |e^{kh} F(e^{kh})| &\leq hC \sum_{k=N_e+1}^{\infty} e^{kh} e^{-\beta e^{kh}} \\
&\leq C \int_{N_e h}^{\infty} e^x e^{-\beta e^x} dx \\
&= \frac{C}{\beta} \exp(-\beta \exp(N_e h))
\end{aligned}$$

where the second inequality can be thought of as bounding the area of inscribed rectangles by the integral. Hence, the error statement for Theorem 3.8 takes the form

$$\left| \int_0^\infty F(u) du - h \sum_{k=-M}^{N_e} e^{kh} F(e^{kh}) \right|$$

$$\leq h \sum_{k=M+1}^{\infty} |e^{-kh} F(e^{-kh})| + h \sum_{k=N_e+1}^{\infty} |e^{kh} F(e^{kh})| + |\eta(F)|$$

$$\leq \frac{C}{\alpha} e^{-\alpha Mh} + \frac{C}{\beta} e^{-\beta e^{N_e h}} + 2N(F, D_W) e^{-2\pi d/h}.$$

The error bound then reads, with N_e in (3.36),

$$\left| \int_0^\infty F(u) du - h \sum_{k=-M}^{N_e} e^{kh} F(e^{kh}) \right|$$

$$\leq K_7 \exp\left(-(2\pi d\alpha M)^{1/2}\right) \qquad (3.37)$$

where

$$K_7 \equiv \left(\frac{C}{\alpha} + \frac{C}{\beta} + 2N(F, D_W)\right).$$

Example 3.11 To illustrate the above development, consider the calculation of Euler's Γ function

$$\int_0^\infty u^{\nu-1} \exp(-u) du = \Gamma(\nu) \qquad (3.38)$$

at $\nu = 3/2$, for example, since in this case the exact value of the integral is $\sqrt{\pi}/2$. Again, the integrand is analytic in the whole of the right half plane, so $d = \pi/2$. The integrand is $\mathcal{O}(\sqrt{u})$ as $u \to 0^+$ so that $\alpha = 3/2$, which (when combined with (3.32)) gives $h = \pi\sqrt{2/(3M)}$. Further, the integrand satisfies (3.35) with $\beta = 1$ so that

$$N_e = \left[\left|\frac{1}{h} \ell n \left(\frac{3}{2} Mh\right) + 1\right|\right].$$

In Table 3.3 the predicted asymptotic convergence rate is given by

$$CR = \exp(-\pi\sqrt{3M/2})$$

and the absolute error is given by

$$AE = \left|\frac{\sqrt{\pi}}{2} - h \sum_{k=-M}^{N_e} e^{3kh/2} \exp(-(e^{kh}))\right|.$$

M	N_e	$h = \pi\sqrt{2/(3M)}$	AE	CR	NOC
4	2	1.283	$.394 \times 10^{-2}$	$.455 \times 10^{-3}$	8.66
8	3	0.907	$.925 \times 10^{-4}$	$.188 \times 10^{-4}$	4.92
16	5	0.641	$.431 \times 10^{-6}$	$.207 \times 10^{-6}$	2.08
32	7	0.453	$.125 \times 10^{-7}$	$.353 \times 10^{-9}$	35.5

Table 3.3 Sinc quadrature using the reduced nodal selection (3.36).

Again, as in Table 3.2, an estimate of the numerical order constant NOC is displayed in Table 3.3 for a sequence of values of M. The column headed N_e has entries that are quite a bit smaller than would have been obtained using the selection in (3.31) (the latter are $3, 6, 12, 24, 42$, respectively). The choice of N in (3.31) does yield the same absolute error, but a large number of negligible contributions are being introduced into the approximation.

∎

As pointed out, a convenient method for the selection of d for the $z = \ell n(w)$ map is the value of d defining the largest sector in the right half plane in which F is analytic. There are occasions where this rule of thumb may lead to confusing results. An exemplary situation of this confusion is provided by the numerical computation of Fourier transforms [3]. For example, an integral that frequently arises is the Fourier sine transform

$$\mathcal{F}(g)(\lambda) = \int_0^\infty g(u)\sin(\lambda u)du, \quad \lambda > 0. \tag{3.39}$$

If g is analytic in the right half plane $(d = \pi/2)$ and satisfies the bound in (3.30) or (3.35), then it may be hastily concluded that the rule (3.33) will do an accurate job of approximating (3.39). There is, however, a difficulty that arises from the assumed bound on the integral in (3.29) with respect to realizing the rate of convergence in (3.24). To see the source of this problem, write

$$\int_r^R |g(\rho e^{id})\sin(\lambda \rho e^{id})|d\rho$$

$$\geq \int_r^R |g(\rho e^{id})|\sinh(\lambda \rho \sin(d))d\rho \tag{3.40}$$

so that, unless

$$|g(\rho e^{id})| = \mathcal{O}(e^{-\rho\gamma}) \tag{3.41}$$

for $\gamma > \lambda(\sin(d))$, one cannot expect the integral in (3.40) to exist. There are a number of approaches to numerically approximate Fourier (or oscillatory) integrals, but the problem is, in general, a very difficult one. The excellent survey of available methods in [1] gives the reader a clear idea of just how onerous the problem can be.

The difficulty with the assumed bound on the integral in (3.40) is that, although $\sin(\lambda u)$ is bounded for real values of its argument, it grows exponentially in the right half plane. If the requirement in (3.29) requires a growth restriction only in a domain in the right half plane with bounded imaginary part, then this exponential growth can be potentially controlled. Indeed, a method based on Theorem 3.8 that has proved a useful tool for the numerical calculation of oscillatory integrals such as (3.39) selects an appropriate domain and map ϕ for the problem at hand. Towards this goal define the domain

$$D_B \equiv \{w \in \mathbb{C} : |\arg(\sinh(w))| < d \leq \pi/2\} \qquad (3.42)$$

which is conformally mapped onto the strip D_S by

$$z = \phi(w) = \ell n(\sinh(w)). \qquad (3.43)$$

This map is the composition of $z = \phi_2(\eta) = \ell n(\eta)$ and $\eta = \phi_1(w) = \sinh(w)$. Since the former has been described, the latter is studied here. Note that

$$\phi_1^{-1}(\eta) = \sinh^{-1}(\eta) = \ell n\left(\eta + \sqrt{\eta^2 + 1}\right).$$

See Figure 3.7 for the domains involved. To gain a feel for these domains, put

$$\begin{aligned}\mu + i\nu = \eta &= \sinh(u + iv) \\ &= \sinh(u)\cos(v) + i\cosh(u)\sin(v).\end{aligned}$$

Upon setting

$$\tan^{-1}\left(\frac{\nu}{\mu}\right) = \arg(\sinh(w)) = \pm d \qquad (3.44)$$

the boundary of D_B is given by

$$\tan(v) = \pm \tan(d)\tanh(u).$$

To verify that this maps onto the boundary of D_W, let $w = u + iv$ be on ∂D_B. Then from (3.44)

$$\nu = \pm \tan(d)\mu$$

which are the two lines that form ∂D_W.

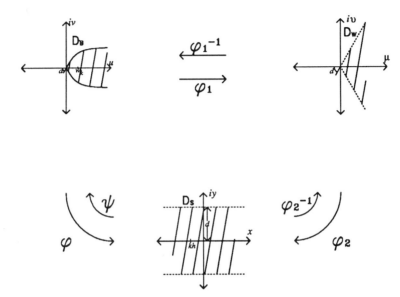

Figure 3.7 The domains D_B, D_W, and D_S.

To see why the $ln(\sinh(w))$ quadrature is better suited for (3.39) than is the $ln(w)$ quadrature, one needs to analyze bounding the integral

$$N(F, D_B) \equiv \lim_{\gamma \to \partial D_B} \int_\gamma |F(w)dw|.$$

As shown above, the upper boundary of D_S (the line $x+di$) is carried onto the curve

$$\gamma^+ \equiv \{(u,v) : \tan(v) = \tan(d)\tanh(u)\}$$

which is the arc of ∂D_B in the upper half plane in Figure 3.5. Thus for the integral in (3.39)

$$\int_{\gamma^+} |g(w)\sin(\lambda w)dw| \leq \cosh(\lambda d) \int_{\gamma^+} |g(w)dw|.$$

Notice that this is a far less restrictive condition on g than is (3.41), for it requires only that g be integrable on an arc asymptotic to $u+id$,

$u \to \infty$. The factor $\cosh(\lambda d)$ can be combined with the bound on the right-hand side of (3.23) to yield the controllable error estimate

$$\eta(F) \simeq \exp(-2\pi d/h) \cosh(\lambda d)$$

if it is assumed that g satisfies the hypotheses of Theorem 3.7. In Example 3.13 the performance of this rule is illustrated, but the form of Theorem 3.8 for the map $\phi(w) = \ln(\sinh(w))$ is first recorded here for future reference.

Theorem 3.12 *Let $F \in B(D_B)$ and let $z = \phi(w) = \ln(\sinh(w))$. Let $\psi(z) = \phi^{-1}(z) = \ln\left[e^z + \sqrt{e^{2z} + 1}\right]$ and $\Gamma = \psi(\mathbb{R}) = (0, \infty)$. Further assume that there are positive constants α, β, and C so that*

$$|F(u)| \leq C \begin{cases} u^{\alpha - 1}, & u \in (0, \ln(1 + \sqrt{2})) \\ \exp(-\beta u), & u \in [\ln(1 + \sqrt{2}), \infty). \end{cases} \quad (3.45)$$

Make the selections

$$N = \left[\left|\frac{\alpha}{\beta} M + 1\right|\right]$$

and

$$h = \left(\frac{2\pi d}{\alpha M}\right)^{1/2} \leq \frac{2\pi d}{\ln(2)}.$$

Then

$$\left|\int_0^\infty F(u)\,du - h \sum_{k=-M}^{N} \frac{F(u_k)}{\phi'(u_k)}\right| = \mathcal{O}\left(\exp\left(-(2\pi d \alpha M)^{1/2}\right)\right) \quad (3.46)$$

where the nodes are

$$u_k = \psi(kh) = \ln\left[e^{kh} + \sqrt{e^{2kh} + 1}\right]$$

and the weights are given by

$$\frac{1}{\phi'(u_k)} = \frac{e^{kh}}{\sqrt{e^{2kh} + 1}}.$$

∎

Notice that the bound in (3.25) takes the form

$$|F(u)| \leq C \begin{cases} \sinh^{\alpha}(u)\coth(u), & u \in \left(0, \ell n(1+\sqrt{2})\right) \\ \sinh^{-\beta}(u)\coth(u), & u \in \left[\ell n(1+\sqrt{2}), \infty\right). \end{cases}$$

The equivalent condition (3.45) is a little easier to check. Three members of the translated sinc basis on the interval $(0, \infty)$ using the map $\phi(u) = \ell n(\sinh(u))$ are shown in Figure 3.8. As the figure illustrates, the nodes u_k are (asymptotically) equispaced for large k.

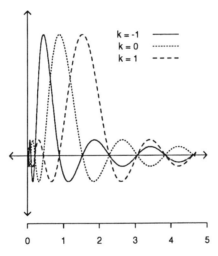

Figure 3.8 The sinc base $(S(k,h) \circ \ell n(\sinh(u)))$ on the interval $(0, \infty)$ for $k = -1, 0, 1, h = \pi/4$.

Example 3.13 The calculation of the Fourier integral

$$\int_0^{\infty} e^{-u} \sin(\lambda u) du = \frac{\lambda}{\lambda^2 + 1} \tag{3.47}$$

can, for moderate λ, be handled by Theorem 3.12. Using the value $d = \pi/2$ with $\alpha = 2$ gives $h = \pi/\sqrt{2M}$ and since $\beta = 1$, $N = 2M$. In Table 3.4 the predicted asymptotic convergence rate is, by (3.33),

$$CR = \exp(-\pi\sqrt{2M}).$$

$\lambda = 1$				
$M = N/2$	$h = \pi/\sqrt{2M}$	AE	CR	NOC
4	1.111	$.397 \times 10^{-3}$	$.138 \times 10^{-3}$	2.87
8	0.785	$.715 \times 10^{-5}$	$.349 \times 10^{-5}$	2.05
16	0.555	$.330 \times 10^{-7}$	$.191 \times 10^{-7}$	1.73
32	0.393	$.154 \times 10^{-10}$	$.122 \times 10^{-10}$	1.27

$\lambda = 5$				
$M = N/2$	$h = \pi/\sqrt{2M}$	AE	CR	NOC
4	1.111	$.677 \times 10^{-1}$	$.138 \times 10^{-3}$	489
8	0.785	$.315 \times 10^{-2}$	$.349 \times 10^{-5}$	904
16	0.555	$.241 \times 10^{-4}$	$.191 \times 10^{-7}$	1259
32	0.393	$.149 \times 10^{-7}$	$.122 \times 10^{-10}$	1227

$\lambda = 10$				
$M = N/2$	$h = \pi/\sqrt{2M}$	AE	CR	NOC
4	1.111	$.537 \times 10^{-1}$	$.138 \times 10^{-3}$	388
8	0.785	$.157 \times 10^{-1}$	$.349 \times 10^{-5}$	4513
16	0.555	$.624 \times 10^{-2}$	$.191 \times 10^{-7}$	32586
32	0.393	$.261 \times 10^{-4}$	$.122 \times 10^{-10}$	2147349

Table 3.4 Sinc quadrature for the Fourier integral (3.47) with the $ln(\sinh(w))$ map.

Furthermore, the absolute error in Table 3.4 is

$$AE = \left| \int_0^\infty e^{-u}\sin(\lambda u)du - h\sum_{k=-M}^{M} \frac{e^{kh}}{\sqrt{e^{2kh}+1}} F(u_k) \right|.$$

Notice that the final column of Table 3.4 is a bit large for $\lambda = 5$ and $\lambda = 10$ compared to the numerical order constant (NOC) from the case $\lambda = 1$. The source of this growth can be seen from an inspection of the approximation in (3.23) (note that $\cosh(5\pi/2) = 1288$ and $\cosh(10\pi/2) = 3317812$).

To fine tune the mesh selection a little, replace the error bound on the right-hand side of (3.23) by the approximation

$$\eta(F) \simeq \exp(-2\pi d/h)\cosh(\lambda d)$$

and now equate this integral error to the desired accuracy of the quadrature, say $10^{-\delta}$, where δ digits is the requested accuracy. Solving for h leads to the mesh selection

$$h_\delta = 2\pi^2/(2\delta\, \ell n(10) + \pi\lambda).$$

Now use the error of truncation in (3.7) to define

$$M_\delta = [|\delta\, \ell n(10)/(\alpha h_\delta) + 1|]$$

and finally, to balance the error $N_\delta = [|(\alpha/\beta)M_\delta + 1|]$. In Table 3.5 these are the selections made in the calculation of the above Fourier integral.

Notice that the corresponding lines in the two tables for $\lambda = 1$ are roughly the same (the mesh h and the number of evaluation points M and M_δ). This is due to the almost negligible contribution of the term $\cosh(1)$. The same comparison for $\lambda = 5$ shows that the results are a little bit better in the case of the fine-tuned mesh selection in Table 3.5. The last line in Table 3.5 shows that the error is 10^{-7} using 61 points, whereas the last line in Table 3.4 shows that 97 points were used to obtain this accuracy. The results corresponding to $\lambda = 10$ are a little more dramatic. In Table 3.5 an accuracy of 10^{-6} is obtained using 79 points, whereas an accuracy of only 10^{-4} was obtained in Table 3.4 using 97 points. The bit of extra work to obtain the mesh selection h_δ seems worth it. The only difference in the calculations to obtain the results of the two tables is in the definition of the mesh size h or h_δ.

		$\lambda = 1$	
δ	$M_\delta = N_\delta/2$	h_δ	AE
3	3	1.164	$.427 \times 10^{-3}$
5	8	0.754	$.248 \times 10^{-5}$
7	15	0.558	$.352 \times 10^{-7}$
10	29	0.401	$.178 \times 10^{-10}$

		$\lambda = 5$	
δ	$M_\delta = N_\delta/2$	h_δ	AE
1	2	0.972	$.964 \times 10^{-2}$
2	3	0.792	$.447 \times 10^{-2}$
4	8	0.578	$.844 \times 10^{-4}$
7	20	0.412	$.673 \times 10^{-7}$

		$\lambda = 10$	
δ	$M_\delta = N_\delta/2$	h_δ	AE
3	8	0.436	$.315 \times 10^{-2}$
5	16	0.363	$.340 \times 10^{-4}$
7	26	0.310	$.372 \times 10^{-6}$
10	46	0.255	$.240 \times 10^{-9}$

Table 3.5 Sinc quadrature for the Fourier integral (3.47) with the fine-tuned selections h_δ and M_δ.

As λ increases, the above method will require an excessive number of function evaluations for accurate computations, as it does with other techniques. If only moderate accuracy is required, then asymptotic methods may meet one's needs.

∎

The error in the trapezoidal rule in Theorem 2.21 for functions defined on the entire real line required the exponential decrease of the integrand at plus and minus infinity (2.35). It is advantageous to have a map ϕ that takes $\mathbb{R}$ to $\mathbb{R}$ and preserves the error bounds in Theorem 3.4 and Theorem 3.8 while requiring only algebraic decay at infinity. This may be due to the rational decay of the integrand or due to a certain inconvenience in converting an integrand to an integral over the semi-axis $(0, \infty)$. Although there is no unique choice for the map ϕ, the resulting quadrature rule has a particularly simple form (expression for the weights and nodes) if one selects $\phi(w) = \sinh^{-1}(w)$.

The domain

$$D_H \equiv \left\{ w \in \mathbb{C} : w = u + iv, \; \frac{v^2}{\sin^2 d} - \frac{u^2}{\cos^2 d} < 1, d \leq \pi/2 \right\} \quad (3.48)$$

is conformally mapped onto the strip D_S by

$$z = \phi(w) = \sinh^{-1}(w) = \ell n \left(w + \sqrt{w^2 + 1} \right).$$

See Figure 3.9 for the domains involved. Let $z = x \pm id$ on ∂D_S so that

$$\begin{aligned} u + iv = w &= \sinh(z) \\ &= \sinh(x)\cos(d) \pm i\cosh(x)\sin(d) \end{aligned}$$

and, upon eliminating x, one obtains the equation of the hyperbola

$$\frac{v^2}{\sin^2(d)} - \frac{u^2}{\cos^2(d)} = 1.$$

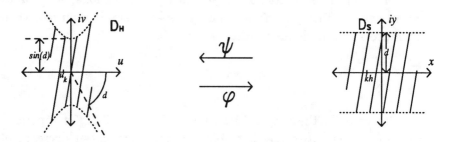

Figure 3.9 The conformally equivalent domains D_H and D_S.

Theorem 3.14 *Let $F \in B(D_H)$ and let $z = \phi(w) = \sinh^{-1}(w)$. Further assume that there are positive constants α, β, and C so that*

$$|F(u)| \leq C \begin{cases} (1+|u|)^{-\alpha-1}, & u \in (-\infty, 0) \\ (1+u)^{-\beta-1}, & u \in [0, \infty). \end{cases}$$

Make the selections

$$N = \left[\left|\frac{\alpha}{\beta} M + 1\right|\right]$$

and

$$h = \left(\frac{2\pi d}{\alpha M}\right)^{1/2} \leq \frac{2\pi d}{\ln 2}.$$

Then

$$\left|\int_{-\infty}^{\infty} F(u)du - h \sum_{k=-M}^{N} \frac{F(u_k)}{\phi'(u_k)}\right| \qquad (3.49)$$

$$= \mathcal{O}\left(\exp\left(-(2\pi d\alpha M)^{1/2}\right)\right)$$

where the nodes u_k and the weights $1/\phi'(u_k)$ are

$$u_k = \sinh(kh), \quad \frac{1}{\phi'(u_k)} = \cosh(kh).$$

∎

The calculation of the nodes and weights in Theorem 3.14 is facilitated by using the identity $\cosh(\sinh^{-1}(u)) = (1+u^2)^{1/2}$. Exercise 3.4 shows that the assumed bound on the integrand in Theorem 3.14 is equivalent to the condition (3.25), which reads

$$|F(u)| \le C \begin{cases} \frac{(|u|+\sqrt{u^2+1})^{-\alpha}}{\cosh(\sinh^{-1}(u))}, & u \in (-\infty, 0) \\ \frac{(u+\sqrt{u^2+1})^{-\beta}}{\cosh(\sinh^{-1}(u))}, & u \in [0, \infty). \end{cases}$$

Three members of the sinc basis on the real line, $\mathbb{R}$, with the map $\phi(u) = \sinh^{-1}(u)$ are shown in Figure 3.10.

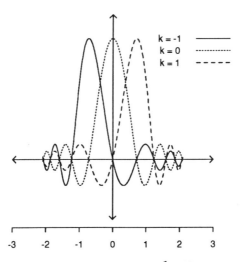

Figure 3.10 The sinc base $(S(k,h) \circ \sinh^{-1}(u))$ on the real line for $k = -1, 0, 1, h = \pi/4$.

Example 3.15 A direct application of the trapezoidal rule to

$$\int_{-\infty}^{\infty} \frac{du}{u^4 + a^4} = \frac{\pi}{\sqrt{2}a^3} \tag{3.50}$$

where a is a positive constant, does not converge exponentially due to the rational behavior of the integrand. This integral may be dealt with by using the fact that the integrand is even, doubling the value of the integral on $(0, \infty)$ and using the $\ell n(w)$ map. The numerical results from this method are listed under the column headed $AE(\ell n)$

in Table 3.6. Exercise 3.6 spells out the parameter choices that give $h = \pi/\sqrt{2M}$ and explain the seemingly inferior performance of the $ln(w)$ versus $\sinh^{-1}(w)$ rule. For the $\sinh^{-1}(w)$ rule the previous theorem gives $\alpha = \beta = 3$ so that $M = N$, and $h = \sqrt{2\pi d/(3M)}$. The only question here is the selection of d, the opening in the hyperbolic rectangle in Figure 3.9. The function $1/(u^4 + a^4)$ has poles at $a\omega$, where ω is a fourth root of unity ($\omega = (1+i)/\sqrt{2}$ in the first quadrant). In the hyperbolic rectangle this corresponds to an angle $d = \pi/4 = \sin^{-1}(1/\sqrt{2})$ (for all a). The absolute error is

$$AE = \left| \frac{\pi}{\sqrt{2}a^3} - h \sum_{k=-M}^{M} \cosh(kh) F(u_k) \right|.$$

M	$h = \pi/(\sqrt{6M})$	AE	$AE(ln)$	$h = \pi/\sqrt{2M}$
4	0.641	$.203 \times 10^{-1}$	$.371 \times 10^{-1}$	1.111
8	0.453	$.216 \times 10^{-2}$	$.438 \times 10^{-2}$	0.785
16	0.321	$.529 \times 10^{-5}$	$.345 \times 10^{-3}$	0.555
32	0.227	$.195 \times 10^{-7}$	$.713 \times 10^{-5}$	0.393

Table 3.6 The $\sinh^{-1}(w)$ rule (3.49) versus the $ln(w)$ rule (3.33) for (3.50) with $a = 1$.

∎

In Table 3.7 the nodes u_k and the weights $1/\phi'(u_k)$ for the quadrature rule in Theorem 3.8 are listed. These are the same nodes that are used for the interpolation formula (3.15). Conditions that bound the right-hand side of (3.14) are itemized for each of the maps discussed in this section. This is the assumed bound for interpolation. The bound needed to satisfy (3.25) for quadrature is the same. The condition on the integrand F is obtained by multiplying the entry in the last column of the top portion of the table by the corresponding entry in the second or third column in the lower portion of the table. The intervals Γ_a and Γ_b, on which the bound is required to hold, are also listed.

(a,b)	$\phi(u)$	u_k	$\frac{1}{\phi'(u)}$
(a,b)	$\ln\left(\frac{u-a}{b-u}\right)$	$\frac{be^{kh}+a}{e^{kh}+1}$	$\frac{(u-a)(b-u)}{(b-a)}$
$(0,1)$	$\ln\left(\frac{u}{1-u}\right)$	$\frac{e^{kh}}{e^{kh}+1}$	$u(1-u)$
$(0,\infty)$	$\ln(u)$	e^{kh}	u
$(0,\infty)$	$\ln(\sinh(u))$	$\ln\left[e^{kh}+\sqrt{e^{2kh}+1}\right]$	$\tanh(u)$
$(-\infty,\infty)$	u	kh	1
$(-\infty,\infty)$	$\sinh^{-1}(u)$	$\sinh(kh)$	$(1+u^2)^{1/2}$

(a,b)	$e^{-\alpha|\phi(u)|}$	$e^{-\beta|\phi(u)|}$	Γ_a	Γ_b
(a,b)	$(u-a)^\alpha$	$(b-u)^\beta$	$\left(a,\frac{a+b}{2}\right)$	$\left[\frac{a+b}{2},b\right)$
$(0,1)$	u^α	$(1-u)^\beta$	$\left(0,\frac{1}{2}\right)$	$\left[\frac{1}{2},1\right)$
$(0,\infty)$	u^α	$u^{-\beta}$	$(0,1)$	$[1,\infty)$
$(0,\infty)$	u^α	$e^{-\beta u}$	$(0,u_0)^*$	$[u_0,\infty)^*$
$(-\infty,\infty)$	$e^{-\alpha|u|}$	$e^{-\beta u}$	$(-\infty,0)$	$[0,\infty)$
$(-\infty,\infty)$	$(1+|u|)^{-\alpha}$	$(1+u)^{-\beta}$	$(-\infty,0)$	$[0,\infty)$

$*u_0 = \ln(1+\sqrt{2}) = \phi^{-1}(0)$

Table 3.7 The nodes, weights, bounds and intervals for interpolation (Theorem 3.4) and quadrature (Theorem 3.8).

3.3 Differentiation on Γ

The error in differentiation on Γ is developed much like the quadrature error. The following theorems from [4] give these results. A weight function g is included to assist in satisfying the hypotheses.

Theorem 3.16 *Let $\phi' F/g \in B(D)$ and $h > 0$. Let ϕ be a one-to-one conformal map of the domain D onto D_S. Let $\psi = \phi^{-1}$ and $\Gamma = \psi(\mathbb{R})$. Assume there exists a constant K such that for all $\xi \in \Gamma$*

$$\left| \frac{d^m}{d\xi^m} \left[\frac{g(\xi) \sin(\pi\phi(\xi)/h)}{2\pi i (\phi(w) - \phi(\xi))} \right] \right|_{w \in \partial D} \le K h^{-m}$$

for $m = 0, 1, \ldots, n$. For all $\xi \in \Gamma$ let

$$\frac{d^m}{d\xi^m} [g(\xi)\epsilon(\phi' F/g)(\xi)] = \frac{d^m F(\xi)}{d\xi^m}$$

$$- \sum_{k=-\infty}^{\infty} \frac{F(w_k)}{g(w_k)} \frac{d^m}{d\xi^m} \left[g(\xi) \operatorname{sinc}\left(\frac{\phi(\xi) - kh}{h} \right) \right]. \quad (3.51)$$

Then for all $m = 0, 1, \ldots, n$

$$\left\| \frac{d^m}{d\xi^m} [g\epsilon(\phi' F/g)] \right\|_\infty \le \frac{K N(\phi F'/g, D)}{h^m \sinh(\pi d/h)}. \quad (3.52)$$

Proof Apply Theorem 3.2 to the function $\phi' F/g$, multiply both sides by $g(\xi)$, and differentiate m times to arrive at

$$\frac{d^m}{d\xi^m}\left[g(x)\epsilon\left(\frac{\phi' F}{g}\right)(\xi)\right]$$

$$= \frac{d^m F(\xi)}{d\xi^m} - \sum_{k=-\infty}^{\infty} \frac{F(w_k)}{g(w_k)} \frac{d^m}{d\xi^m}\left[g(\xi)\operatorname{sinc}\left(\frac{\phi(\xi)-kh}{h}\right)\right]$$

$$= \frac{d^m}{d\xi^m}\left[\frac{g(\xi)\sin(\pi\phi(\xi)/h)}{2\pi i}\right] \lim_{\gamma \to \partial D} \int_\gamma \frac{[F(w)\phi'(w)/g(w)]dw}{(\phi(w)-\phi(\xi))\sin(\pi\phi(w)/h)}$$

$$= \lim_{\gamma \to \partial D} \int_\gamma \frac{d^m}{d\xi^m}\left[\frac{g(\xi)\sin(\pi\phi(\xi)/h)}{2\pi i(\phi(w)-\phi(\xi))}\right] \frac{[F(w)\phi'(w)/g(w)]}{\sin(\pi\phi(w)/h)} dw.$$

Thus
$$\left|\frac{d^m}{d\xi^m}\left[g(x)\epsilon\left(\frac{\phi'F}{g}\right)(\xi)\right]\right| \leq \frac{K}{h^m\sinh(\pi d/h)}\lim_{\gamma\to\partial D}\int_\gamma\left|\frac{F(w)\phi'(w)}{g(w)}\right|dw$$

and (3.52) results. ∎

Theorem 3.17 *Let $\phi'F/g \in B(D)$ and $h > 0$. Let ϕ be a one-to-one conformal map of the domain D onto D_S. Let $\psi = \phi^{-1}$ and $\Gamma = \psi(\mathbb{R})$. Further assume that there exist constants K and L, such that for all $\xi \in \Gamma$*

$$\left|\frac{d^m}{d\xi^m}\left[\frac{g(\xi)\sin(\pi\phi(\xi)/h)}{2\pi i(\phi(w)-\phi(\xi))}\right]\right|\bigg|_{w\in\partial D} \leq Kh^{-m}$$

and

$$\left|\frac{d^m}{d\xi^m}\left[g(\xi)\operatorname{sinc}\left(\frac{\phi(\xi)-kh}{h}\right)\right]\right| \leq Lh^{-m}$$

for all $m = 0, 1, \ldots, n$. Finally, assume that there are positive constants α, β, and C so that

$$\left|\frac{F(\xi)}{g(\xi)}\right| \leq C\begin{cases}\exp(-\alpha|\phi(\xi)|), & \xi \in \Gamma_a \\ \exp(-\beta|\phi(\xi)|), & \xi \in \Gamma_b\end{cases}$$

where Γ_q ($q = a, b$) are given in (3.8) and (3.9), respectively. If the selections

$$N = \left[\left|\frac{\alpha}{\beta}M + 1\right|\right]$$

and

$$h = \left(\frac{\pi d}{\alpha M}\right)^{1/2} \leq \frac{2\pi d}{\ln 2}$$

are made, then for all $\xi \in \Gamma$ let

$$\frac{d^m}{d\xi^m}[g(\xi)\epsilon_{M,N}(\phi'F/g)(\xi)]$$
$$= \frac{d^m F(\xi)}{d\xi^m} - \sum_{k=-M}^{N}\frac{F(w_k)}{g(w_k)}\frac{d^m}{d\xi^m}\left[g(\xi)\operatorname{sinc}\left(\frac{\phi(\xi)-kh}{h}\right)\right].$$

Then

$$\left\| \frac{d^m}{d\xi^m} [g\epsilon_{M,N}(\phi'F/g)] \right\|_\infty \tag{3.53}$$

$$\leq K_8 M^{(m+1)/2} \exp\left(-(\pi d\alpha M)^{1/2}\right)$$

for all $m = 0, 1, \ldots, n$.

Proof From Theorem 3.16, since $h \leq 2\pi d/(\ell n 2)$,

$$\left| \frac{d^m}{d\xi^m} g(\xi)\epsilon(\phi'F/g)(\xi) \right| \leq \frac{4KN(\phi'F/g, D)}{h^m} e^{-\pi d/h}.$$

Thus

$$\left| \frac{d^m}{d\xi^m} [g(\xi)\epsilon_{M,N}(\phi'F/g)(\xi)] \right|$$

$$\leq \sum_{k=M+1}^{\infty} \left| \frac{F(w_{-k})}{g(w_{-k})} \right| \left| \frac{d^m}{d\xi^m} \left[g(\xi) \operatorname{sinc}\left(\frac{\phi(\xi) - kh}{h}\right) \right] \right|$$

$$+ \sum_{k=N+1}^{\infty} \left| \frac{F(w_k)}{g(x_k)} \right| \left| \frac{d^m}{d\xi^m} \left[g(\xi) \operatorname{sinc}\left(\frac{\phi(\xi) - kh}{h}\right) \right] \right|$$

$$+ \frac{4KN(\phi'F/g, D)}{h^m} e^{-\pi d/h}$$

$$\leq \frac{CL}{\alpha h^{m+1}} e^{-\alpha Mh} + \frac{CL}{\beta h^{m+1}} e^{-\beta Nh} + \frac{4K}{h^m} N(\phi'F/g, D) e^{-\pi d/h}.$$

Making the selections h and N gives (3.53), where

$$K_8 \equiv \left[CL\left(\frac{1}{\alpha} + \frac{1}{\beta}\right) + 4L\left(\frac{\pi d}{\alpha}\right)^{1/2} N(\phi'F/g, D) \right] \left(\frac{\alpha}{\pi d}\right)^{(m+1)/2}.$$

∎

The approximation to the m-th derivative of F in Theorem 3.17 is simply an m-th derivative of $C_{M,N}(F/g, h, \phi)/g$ in (3.15). The weight function g is chosen relative to the order of the derivative that is to be approximated. For instance, if one wishes to approximate an m-th derivative, then the choice

$$g(u) = \frac{1}{(\phi'(u))^m} \tag{3.54}$$

often suffices. For $m = 1$, this choice works fine for the approximation of the first derivative in the next example.

Example 3.18 The function

$$F(u) = u^{3/2}(1-u)ln(u) \tag{3.55}$$

satisfies the hypotheses of Theorem 3.2 with $\alpha = 1/2$ and $\beta = 1$. To see how to obtain these selections, put

$$F(u)/g(u) = u^{1/2}ln(u)$$

where, as recommended above, $g(u) = 1/\phi'(u) = u(1-u)$. For all positive ϵ

$$\lim_{u \to 0} \frac{u^{1/2}ln(u)}{u^{1/2-\epsilon}} = 0$$

so $\alpha = 1/2$. In a neighborhood of $u = 1$, say $u \in (1/2, 1]$,

$$\left|u^{1/2}ln(u)\right| \le |ln(u)| \le 2ln(2)(1-u)$$

so $\beta = 1$. Selecting $h = \pi/\sqrt{M}$ and $N = [|1/2\,M + 1|]$, where $d = \pi/2$, the error in approximating F' by the sinc expansion

$$\frac{d}{du}(gC_{M,N}(F/g, h, \phi))(u) = \sum_{j=-M}^{N} \frac{F(u_j)}{g(u_j)} (gS(j,h) \circ \phi)'(u)$$

yields a convergence rate (CR), which is $\mathcal{O}((M\exp(-\pi\sqrt{M}/2)))$. In order to preview a topic in the next chapter, define

$$\delta_{jk}^{(1)} \equiv h\frac{d}{d\phi}[S(j,h) \circ \phi(u)]\bigg|_{u=u_k} = \begin{cases} 0, & j = k \\ \frac{(-1)^{k-j}}{k-j}, & j \ne k \end{cases} \tag{3.56}$$

so that the above approximation at the sinc nodes u_k takes the form

$$\frac{d}{du}(gC_{M,N}(F/g, h, \phi))(u_k) = \sum_{j=-M}^{N} \left(\frac{\delta_{jk}^{(1)}}{h} + \delta_{jk}^{(0)} g'(u_j)\right) \frac{F(u_j)}{g(u_j)}.$$

To get a feel for the error in the approximation, define the error on the sinc grid by

$$\|E_S(h)\| = \max_{-M \le k \le N} \left| F'(u_k) - \frac{d}{du}(gC_{M,N}(F/g, h, \phi))(u_k) \right|.$$

The final two columns of Table 3.8 compare the error on this sinc grid and the convergence rate CR. The ratio $\|E_S(h)\|/CR$ is roughly one half for all M. One could also compute the error on a uniform grid but, as in Example 3.6, the error given by CR is a more faithful representation of the error of the method than is the error on the uniform grid.

$M = 2N$	$\|E_S(h)\|$	CR	$\|E_S(h)\|/CR$
4	$.794 \times 10^{-1}$	$.173 \times 10^{-0}$	.459
8	$.469 \times 10^{-1}$	$.941 \times 10^{-1}$	.498
16	$.161 \times 10^{-1}$	$.299 \times 10^{-1}$	.538
32	$.255 \times 10^{-2}$	$.443 \times 10^{-2}$	.577

Table 3.8 Sinc approximation of the derivative of F in (3.55).

Notice that for the given function one has the sinc approximation

$$C_{M,N}(F, h, \phi)(u) = \sum_{j=-M}^{N} F(u_j) S(j, h) \circ \phi(u)$$

which by Theorem 3.5 provides an accurate approximation to $F(u)$. Differentiating this interpolation formula and evaluating at u_k yields the expression

$$\frac{d}{du}(C_{M,N}(F, h, \phi))(u_k) = \sum_{j=-M}^{N} \frac{\delta_{jk}^{(1)}}{h} \phi'(u_j) F(u_j).$$

In the present example this expression, if taken to be an approximation of $F'(u_k)$, is very poor. This is due to the fact that $\phi'(u)$ is unbounded as the nodes approach zero or one but for the given function $F'(1) = 0$. This is the reason for the introduction of the weight g in Theorem 3.17.

■

Exercises

Exercise 3.1 Consider the map $\phi(w) = \ln\left(\frac{w}{1-w}\right)$. Find the conditions analogous to (3.28), (3.29), and (3.30) so that

$$\left| \int_0^1 F(u)\,du - h \sum_{k=-M}^{N} \frac{e^{kh}}{(e^{kh}+1)^2} F\left(\frac{e^{kh}}{e^{kh}+1}\right) \right|$$
$$= \mathcal{O}\left(e^{-(2\pi d\alpha M)^{1/2}}\right).$$

Exercise 3.2 Compute the integral

$$\int_0^\infty \left(\frac{1}{u+1} - e^{-u}\right) \frac{du}{u} = .57721566490153\ldots$$

using the optimal $\alpha = 2$ discussed in Example 3.10. The only alteration to the parameter selections used in that example are a change to $h = \pi/\sqrt{2M}$ and $N = 2M$ used for this exercise. Compare these results with the results in Table 3.2. While the results in this exercise are more accurate than the results in Table 3.2, 50 percent more nodes are being used in the present exercise. The point is that either selection of α defines a mesh size that renders an accurate approximation to the integral using the quadrature rule in Theorem 3.9.

Exercise 3.3 The integral

$$I(\mu) = \int_0^\infty \frac{x^{\mu-1}}{(1+x)^3}\,dx = \frac{1}{2}\Gamma(\mu)\Gamma(3-\mu)$$
$$= \begin{cases} \frac{3}{8}\pi, & \mu = 1/2 \\ \frac{1}{2}, & \mu = 1 \\ \frac{1}{8}\pi, & \mu = 3/2 \end{cases}$$

is singular at the origin in the case that $\mu = 1/2$ and $3/2$. Using (3.33), approximate $I(\mu)$ for each of $\mu = 1/2$, 1, and $3/2$, respectively. For each μ, tabulate your results in a fashion analogous to the display in Table 3.1. Notice that the accuracy obtained for the $I(3/2)$ is better than that obtained for $I(1)$ in spite of the fact that the former has a singular integrand, while the latter is analytic in the closed right half plane.

Exercise 3.4 Using the identity

$$\sinh^{-1}(u) = \ell n(u + \sqrt{u^2 + 1})$$

show that the condition (3.25) may be replaced by the condition listed in Theorem 3.14 for the $\sinh^{-1}(u)$ map.

Exercise 3.5 Compute the integral

$$\int_{-\infty}^{\infty} e^{-x^2} dx = \sqrt{\pi}$$

by converting the integral to one on $(0, \infty)$ and using the $\phi(w) = \ell n(w)$ quadrature rule. Compare these results with the results in Exercise 2.1. It may not always be the best idea to use a map on an integrand [2].

Exercise 3.6 Compute the integral

$$\int_{-\infty}^{\infty} \frac{1}{u^4 + 1} du = \frac{\pi}{\sqrt{2}}$$

by writing the integral as an integral over $(0, \infty)$ and applying the $(\phi(w) = \ell n(w))$ quadrature rule. Show that the appropriate parameter selections are $\alpha = 1$ and $\beta = 3$ so that $h = \pi/\sqrt{2M}$ and $N = M/3$. If the approximation is computed for $M = 4, 8, 16$, and 32, then the results under the column labeled $AE(\ell n)$ in Table 3.6 are obtained. Argue that the results that should be compared for equal work correspond to $M = 6, 12, 24$, and 48. Recompute the integral using these values of M. Your results will match the results under the column labeled AE in Table 3.6.

References

[1] P. J. Davis and P. Rabinowitz, *Methods of Numerical Integration*, Second Ed., Academic Press, Inc., San Diego, 1984.

[2] N. Eggert and J. Lund, "The Trapezoidal Rule for Analytic Functions of Rapid Decrease," *J. Comput. Appl. Math.*, 27 (1989), pages 389–406.

[3] J. Lund, "Sinc Function Quadrature Rules for the Fourier Integral," *Math. Comp.*, 41 (1983), pages 103–113.

[4] L. Lundin and F. Stenger, "Cardinal Type Approximations of a Function and Its Derivatives," *SIAM J. Math. Anal.*, 10 (1979), pages 139–160.

[5] I. P. Natanson, *Constructive Function Theory, Vol. I: Uniform Approximation*, Frederick Ungar Publishing Co., Inc., New York, 1964.

[6] F. Stenger, "Integration Formulae Based on the Trapezoidal Formula," *J. Inst. Maths. Appl.*, 12 (1973), pages 103–114.

[7] F. Stenger, "Approximations via Whittaker's Cardinal Function," *J. Approx. Theory*, 17 (1976), pages 222–240.

Chapter 4

The Sinc-Galerkin Method

The material developed in the previous chapters provides the foundation for the formulation and study of the Sinc-Galerkin method. In this chapter it is specifically applied to the second-order two-point boundary value problem. However, the ideas in this chapter also provide the necessary tools for the Sinc-Galerkin solution of partial differential equations occurring in the last two chapters. Section 4.1 briefly outlines the concept of a Galerkin method as it applies to the two-point boundary value problem. This brief introduction makes it fairly transparent that, from a numerical standpoint, the quadrature rules will play a pervasive role. As such, the necessary inner product approximations, which are based on the quadrature rules in Section 3.2, are spelled out in Section 4.2. The assembly of the discrete system and the convergence of the method are considered in Section 4.3. As in the last chapter, a number of examples are included to highlight and illustrate the method's implementation. Section 4.4 describes the treatment of more general boundary conditions. The final Section 4.5 outlines various analytic issues, including the conditioning of the linear system and optimal parameter selections.

4.1 A Galerkin Scheme

Very accurate solutions to the boundary value problem

$$-u''(x) + q(x)u(x) = f(x), \quad -\infty < x < \infty$$
$$\lim_{x \to \pm\infty} u(x) = 0$$

may be obtained by the Sinc-Galerkin procedure outlined in Example 2.22. In this chapter, the Sinc-Galerkin scheme will be developed for the two-point boundary value problem

$$Lu(x) \equiv -u''(x) + p(x)u'(x) + q(x)u(x) = f(x), \quad a < x < b \tag{4.1}$$
$$u(a) = u(b) = 0.$$

In order to develop this methodology, a Galerkin scheme is reviewed in this section. This material is developed nicely in [5] or [7], wherein the various connections and distinctions between Galerkin and Rayleigh–Ritz are spelled out. In particular, the introduction in [5] provides a nice historical perspective on the development of the methods.

Consider the differential equation $Lu = f$ on a Hilbert space H, where L is given by (4.1) and it is assumed that the solution $u \in H$. Let $\{\chi_i\}_{i=1}^{\infty}$ be a dense linearly independent set in H and let $H^N = \text{span}\{\chi_1, \ldots, \chi_N\}$. Define the approximation $u_N \in H^N$ to the solution $u \in H$ by the linear combination

$$u_N(x) = \sum_{k=1}^{N} c_k \chi_k(x) \tag{4.2}$$

where the coefficients $\{c_k\}$ are to be determined. Galerkin's idea is to require the *residual*

$$R_N(x) \equiv (Lu_N - f)(x)$$

to be orthogonal to each $\chi_j \in H^N$ in the inner product $(\cdot, \cdot)$ on H. That is,

$$(R_N, \chi_j) = 0, \quad 1 \leq j \leq N. \tag{4.3}$$

THE SINC-GALERKIN METHOD

Writing out the system in (4.3) leads to the matrix problem

$$\sum_{k=1}^{N}(L\chi_k,\chi_j)c_k = (f,\chi_j), \quad 1\leq j\leq N \tag{4.4}$$

for the determination of the $\{c_k\}$.

Notice that if the $\{\chi_k\}$ do not satisfy the boundary conditions in (4.1), then one cannot expect the approximate solution to satisfy the boundary conditions. Thus the approach in the Galerkin method is to select the basis functions so that each χ_k satisfies the boundary conditions in (4.1).

The properties of the Galerkin solution (4.2) for the problem (4.1) depend both on the inner product on H and the selection of the set of functions $\{\chi_i\}_{i=1}^{\infty}$. To fix ideas here, define the inner product by

$$(f,g) = \int_a^b f(x)g(x)w(x)dx \tag{4.5}$$

where w is a weight function. In this case, the elements of the matrix on the left-hand side of (4.4) take the form

$$\begin{aligned}(L\chi_k,\chi_j) &= \int_a^b L\chi_k(x)\chi_j(x)w(x)dx \\ &= -\int_a^b \chi_k''(x)\chi_j(x)w(x)dx + \int_a^b p(x)\chi_k'(x)\chi_j(x)w(x)dx \\ &\quad + \int_a^b q(x)\chi_k(x)\chi_j(x)w(x)dx.\end{aligned}$$

The elements of the column vector on the right-hand side of (4.4) are

$$(f,\chi_j) = \int_a^b f(x)\chi_j(x)w(x)dx.$$

Hence, the construction of the discrete system in (4.4) depends on the approximation of the various integrals. For the Sinc-Galerkin method, these integrals can be accurately computed via the quadratures of Section 3.2. This development is from [12] and forms the content of the next section.

4.2 Inner Product Approximations

Assume the interval (a, b) in (4.1) is contained in $\mathbb{R} = (-\infty, \infty)$, and let ϕ be one of the conformal maps of Chapter 3 determined by the interval (a, b). Define the basis functions of the previous section by

$$\chi_j(x) \equiv S(j, h) \circ \phi(x) \tag{4.6}$$

$$= \operatorname{sinc}\left[\frac{\phi(x) - jh}{h}\right], \quad j \in \mathbb{Z}.$$

Throughout this chapter, with the exception of Theorem 4.2, the independent variable will be x so that the dependent variable for the boundary value problems can be designated by u.

The orthogonalization of the residual in (4.3) takes the form

$$(Lu_m - f, S(j, h) \circ \phi) = 0 \tag{4.7}$$

where

$$u_m(x) = \sum_{k=-M}^{N} u_k S(k, h) \circ \phi(x), \quad m = M + N + 1. \tag{4.8}$$

This orthogonalization of the residual is more conveniently treated by first considering the inner products in

$$(Lu - f, S(j, h) \circ \phi) = 0. \tag{4.9}$$

Using the quadrature rule in Theorem 3.8 to approximate each of the integrals arising from (4.9) leads to the same discrete system as the system obtained from (4.7). Although there is no difference in these developments for the homogeneous problem (4.1), having the two alternative points of view is particularly handy in the treatment of the nonhomogeneous problem in Section 4.4 and in the discretization of partial differential equations in Section 5.4 and Chapter 6.

Instead of directly applying the quadrature rule to the inner products, the Sinc-Galerkin method proceeds as follows. In each inner product arising from (4.9) integrate by parts those terms containing derivatives of u to remove the derivatives from u. For example, the

THE SINC-GALERKIN METHOD

inner product containing the first derivative of u reads

$$(pu', S(j,h) \circ \phi) = \int_a^b p(x)u'(x)[S(j,h) \circ \phi(x)]w(x)dx \qquad (4.10)$$

$$= B_{T_1} - \int_a^b u(x)(p[S(j,h) \circ \phi]w)'(x)dx$$

where

$$B_{T_1} = (up[S(j,h) \circ \phi]w)(x)\Big|_a^b. \qquad (4.11)$$

The reason for the integration by parts is twofold. Firstly, the selection of the weight w can be based on criteria such as forcing the boundary term B_{T_1} to vanish. Secondly, u_m can for suitable u_j ($u_m(x_j) = u_j \simeq u(x_j)$, $x_j = \phi^{-1}(jh)$) provide a good approximation of u (Theorem 3.4). However, the derivatives of u_m in (4.8) do not necessarily give accurate approximations to the corresponding derivatives of u near a or b. Recall the discussion following Table 3.8. This difficulty has been removed in the second equality in (4.10).

The inner product containing u'' takes the form

$$(u'', S(j,h) \circ \phi) = \int_a^b u''(x)[S(j,h) \circ \phi(x)]w(x)dx \qquad (4.12)$$

$$= B_{T_2} + \int_a^b u(x)([S(j,h) \circ \phi]w)''(x)dx$$

where

$$B_{T_2} = (u'[S(j,h) \circ \phi]w)(x)\Big|_a^b - (u([S(j,h) \circ \phi]w)')(x)\Big|_a^b. \qquad (4.13)$$

An approach involving only one integration by parts in (4.12), which is more in the spirit of the Rayleigh–Ritz method, is described in [2].

The method of approximation of the integrals in (4.10) and (4.12) begins with the quadrature rule in Theorem 3.7. To record these approximations, it is convenient to collect the notation introduced in (2.47) and (3.56)

$$\delta_{jk}^{(p)} \equiv h^p \frac{d^p}{d\phi^p}[S(j,h) \circ \phi(x)]\Big|_{x=x_k}, \quad p = 0, 1, 2, \ldots$$

where $x_k = \phi^{-1}(kh) = \psi(kh)$. Explicitly, these quantities (for $p = 0, 1, 2$) are given in the following theorem.

Theorem 4.1 *Let ϕ be a conformal one-to-one map of the simply connected domain D onto D_S. Then*

$$\delta_{jk}^{(0)} \equiv [S(j,h) \circ \phi(x)]\Big|_{x=x_k} = \begin{cases} 1, & j=k \\ 0, & j \neq k, \end{cases} \qquad (4.14)$$

$$\delta_{jk}^{(1)} \equiv h\frac{d}{d\phi}[S(j,h) \circ \phi(x)]\Big|_{x=x_k} = \begin{cases} 0, & j=k \\ \frac{(-1)^{k-j}}{k-j}, & j \neq k, \end{cases} \qquad (4.15)$$

and

$$\delta_{jk}^{(2)} \equiv h^2\frac{d^2}{d\phi^2}[S(j,h) \circ \phi(x)]\Big|_{x=x_k} = \begin{cases} \frac{-\pi^2}{3}, & j=k \\ \frac{-2(-1)^{k-j}}{(k-j)^2}, & j \neq k. \end{cases} \qquad (4.16)$$

Proof Each of (4.14)–(4.16) may be seen by noting that

$$\frac{d^n}{d\phi^n}[S(j,h) \circ \phi(x)]\Big|_{x=x_k} = \frac{d^n}{dx^n}\left[\operatorname{sinc}\left(\frac{x-jh}{h}\right)\right]\Big|_{x=kh}$$

and recalling

$$\operatorname{sinc}\left(\frac{x-jh}{h}\right) = \frac{h}{2\pi}\int_{-\pi/h}^{\pi/h} e^{-ixt} e^{ijht} dt.$$

This representation gives the general expressions

$$\frac{d^{4n}}{dx^{4n}}\left[\operatorname{sinc}\left(\frac{x-jh}{h}\right)\right] = \frac{h}{\pi}\int_0^{\pi/h} t^{4n} \cos[t(x-jh)]dt, \qquad (4.17)$$

$$\frac{d^{4n+1}}{dx^{4n+1}}\left[\operatorname{sinc}\left(\frac{x-jh}{h}\right)\right] = \frac{-h}{\pi}\int_0^{\pi/h} t^{4n+1} \sin[t(x-jh)]dt, \qquad (4.18)$$

$$\frac{d^{4n+2}}{dx^{4n+2}}\left[\operatorname{sinc}\left(\frac{x-jh}{h}\right)\right] = \frac{-h}{\pi}\int_0^{\pi/h} t^{4n+2} \cos[t(x-jh)]dt, \qquad (4.19)$$

$$\frac{d^{4n+3}}{dx^{4n+3}}\left[\operatorname{sinc}\left(\frac{x-jh}{h}\right)\right] = \frac{h}{\pi}\int_0^{\pi/h} t^{4n+3} \sin[t(x-jh)]dt. \qquad (4.20)$$

Setting $n=0$ in (4.17)–(4.19) leads to (4.14)–(4.16). ∎

One notes with the representations in (4.17) through (4.20) that there is the potential for defining $\delta_{jk}^{(p)}$ for all integers p by the above integrals. This is true and in Exercise 4.6 each of $\delta_{jk}^{(3)}$ and $\delta_{jk}^{(4)}$ is met via the consideration of a fourth-order boundary value problem [10]. In order to bound the various integrands that arise from the application of Theorem 3.7 to the inner products in (4.10) and (4.12), the following inequalities are useful.

Theorem 4.2 *Let ϕ be a conformal one-to-one map of the simply connected domain D onto D_S. Then for $d, h > 0$,*

$$\left|\frac{S(j,h)\circ\phi(w)}{\sin(\pi\phi(w)/h)}\right|_{w\in\partial D} \leq \frac{h}{\pi d} \equiv C_0(h,d), \qquad (4.21)$$

$$\left|\frac{\frac{d}{d\phi}[S(j,h)\circ\phi(w)]}{\sin(\pi\phi(w)/h)}\right|_{w\in\partial D} \leq \frac{\pi d + h\tanh(\pi d/h)}{\pi d^2 \tanh(\pi d/h)} \equiv C_1(h,d) \quad (4.22)$$

and

$$\left|\frac{\frac{d^2}{d\phi^2}[S(j,h)\circ\phi(w)]}{\sin(\pi\phi(w)/h)}\right|_{w\in\partial D} \leq \frac{2}{d}C_1(h,d) + \frac{\pi}{hd} \equiv C_2(h,d). \quad (4.23)$$

Furthermore, for $p = 0, 1, 2$,

$$C_p(h,d) = 0(h^{1-p}), \quad h \to 0.$$

Proof Upon noting that

$$\left|\frac{\frac{d^n}{d\phi^n}[S(j,h)\circ\phi(w)]}{\sin(\pi\phi(w)/h)}\right|_{w\in\partial D} = \left|\frac{\frac{d^n}{dz^n}[S(j,h)(z)]}{\sin(\pi z/h)}\right|_{z=x\pm id}$$

the inequality in (4.21) is bounded as follows:

$$\left|\frac{S(j,h)\circ\phi(w)}{\sin(\pi\phi(w)/h)}\right|_{w\in\partial D} = \left|\frac{\operatorname{sinc}\left(\frac{z-jh}{h}\right)}{\sin(\pi z/h)}\right|_{z=x\pm id}$$

$$= \left|\frac{h}{\pi(z-jh)}\right|\bigg|_{z=x\pm id}$$

$$= \frac{h}{|\pi(x-jh)\pm i\pi d|}$$

$$\leq \frac{h}{\pi d}.$$

The bounds in (4.22) and (4.23) are proven in a similar fashion using (2.33) and the inequality

$$|\cos[\pi(x\pm id)/h]| \leq \cosh(\pi d/h).$$

∎

It should be noted that for $w \in D$ the calculation

$$\left|\frac{d^n}{d\phi^n}[S(j,h)\circ\phi(w)]\right| = \left|\frac{d^n}{dz^n}[S(j,h)(z)]\right|$$

$$= \frac{h}{\pi}\int_0^{\pi/h} t^n \left\{\begin{array}{ll}\cos[t(z-jh)], & n \text{ even} \\ \sin[t(z-jh)], & n \text{ odd}\end{array}\right\} dt$$

$$\leq \left(\frac{\pi}{h}\right)^{n-1}\frac{\sinh(\pi d/h)}{d}$$

shows that all the derivatives $\frac{d^n}{d\phi^n}[S(j,h)\circ\phi(w)]$ are bounded on D. Thus if $F \in B(D)$, $F\left\{\frac{d^n}{d\phi^n}[S(j,h)\circ\phi]\right\} \in B(D)$. These results provide the tools to prove the following theorems concerning the error in the sinc quadrature applied to the inner products in (4.9).

Theorem 4.3 *Let ϕ be a conformal one-to-one map of the simply connected domain D onto D_S. Assume $\phi(a) = -\infty$, $\phi(b) = +\infty$ and let $x_k = \phi^{-1}(kh)$.*
(a) Let $vw \in B(D)$ for $v = f$ or qu. Then

$$\left|\int_a^b (vw[S(j,h)\circ\phi])(x)dx - h\left(\frac{vw}{\phi'}\right)(x_j)\right|$$

(4.24)

$$\leq \frac{C_0(h,d)}{2} N(vw,D)e^{-\pi d/h}.$$

(b) Let $u(p[S(j,h) \circ \phi]w)' \in B(D)$ and B_{T_1}, given in (4.11), be zero. Then

$$\left| \int_a^b (pu'[S(j,h) \circ \phi]w)(x)dx \right.$$
$$\left. + h \sum_{k=-\infty}^{\infty} (upw)(x_k) \frac{\delta_{jk}^{(1)}}{h} + h \left(\frac{u(pw)'}{\phi'} \right)(x_j) \right| \quad (4.25)$$
$$\leq \frac{1}{2} [C_1(h,d) N(upw\phi', D)$$
$$+ C_0(h,d) N(u(pw)', D)] e^{-\pi d/h}.$$

(c) Let $u([S(j,h) \circ \phi]w)'' \in B(D)$ and B_{T_2}, given in (4.13), be zero. Then

$$\left| \int_a^b (u''[S(j,h) \circ \phi]w)(x)dx - h \sum_{k=-\infty}^{\infty} u(x_k) \left[\frac{\delta_{jk}^{(2)}}{h^2} (\phi'w)(x_k) \right.\right.$$
$$\left.\left. + \frac{\delta_{jk}^{(1)}}{h} \left(\frac{\phi''}{\phi'} w + 2w' \right)(x_k) \right] - h \left(\frac{w''u}{\phi'} \right)(x_j) \right|$$
(4.26)
$$\leq \frac{1}{2} [C_0(h,d) N(uw'', D) + C_1(h,d) N(u[\phi''w + 2\phi'w'], D)$$
$$+ C_2(h,d) N(u(\phi')^2 w, D)] e^{-\pi d/h}.$$

Proof Apply Theorem 3.7 to the integrands in (a), (b), and (c), where in (b) and (c) the integration by parts in (4.10) and (4.12), respectively, is performed first. For example, in the case of (b)

$$\int_a^b (pu'[S(j,h) \circ \phi]w)(x)dx$$
$$= B_{T_1} - \left\{ \int_a^b \left(up \frac{d}{d\phi} [S(j,h) \circ \phi] \phi' w \right)(x) dx \right.$$
$$\left. + \int_a^b (u[S(j,h) \circ \phi](pw)')(x) \right\}$$
$$= -h \sum_{k=-\infty}^{\infty} (upw)(x_k) \frac{\delta_{jk}^{(1)}}{h} - h \left(\frac{u(pw)'}{\phi'} \right)(x_j)$$
$$- \frac{i}{2} \lim_{\gamma \to \partial D} \int_\gamma \frac{F(z)\kappa(\phi,h)(z)}{\sin(\pi\phi(z)/h)} dz - \frac{i}{2} \lim_{\gamma \to \partial D} \int_\gamma \frac{G(z)\kappa(\phi,h)(z)}{\sin(\pi\phi(z)/h)} dz$$

where $F = up\frac{d}{d\phi}[S(j,h)\circ\phi]\phi'w$, $G = u[S(j,h)\circ\phi](pw)'$, and (4.15) was used in the second equality. Taking absolute values and using each of (4.21), (4.22), and (3.24) leads to the bound in (4.25). Each of (a) and (c) is obtained in an analogous fashion.

∎

In order to bound the truncation error of the infinite sums in Theorem 4.3, further assumptions on the functions u, p, and q are required if one is to obtain bounds that converge exponentially with the number of sample nodes used and the mesh size h. The situation here is reminiscent of that in Example 2.10, where, although the infinite sum converges exponentially with the mesh spacing h, the error in truncation converges algebraically with h. The assumptions in the following theorem are analogous to those of the quadrature Theorem 3.8. This is not too surprising when one recalls that what is being approximated are various integral inner products.

Theorem 4.4 *Let ϕ be a conformal one-to-one map of the simply connected domain D onto D_S. Assume $\phi(a) = -\infty$ and $\phi(b) = +\infty$; also let $x_k = \phi^{-1}(kh)$. Further assume that there exist positive constants α, β, and K so that*

$$|F(x)| \leq K \begin{cases} \exp(-\alpha|\phi(x)|), & x \in \Gamma_a \\ \exp(-\beta|\phi(x)|), & x \in \Gamma_b \end{cases} \quad (4.27)$$

where $F = upw$, $u\phi'w$, or $u\left[\frac{\phi''}{\phi'}w + 2w'\right]$ and Γ_q ($q = a,b$) are defined in (3.8) and (3.9), respectively. Make the selections

$$N = \left[\left|\frac{\alpha}{\beta}M + 1\right|\right] \quad (4.28)$$

and

$$h = \left(\frac{\pi d}{\alpha M}\right)^{1/2}. \quad (4.29)$$

(a) Let $vw \in B(D)$ for $v = f$ or qu. Then

$$\left|\int_a^b (vw[S(j,h)\circ\phi])(x)dx - h\left(\frac{vw}{\phi'}\right)(x_j)\right| \quad (4.30)$$

$$\leq L_0 M^{-1/2}\exp(-(\pi d\alpha M)^{1/2})$$

where L_0 is a constant depending on v, w, and d.

(b) Let $u(p[S(j,h) \circ \phi]w)' \in B(D)$ and B_{T_1} in (4.11) be zero. Then

$$\left| \int_a^b (pu'[S(j,h) \circ \phi]w)(x)dx \right.$$
$$\left. + h \sum_{k=-M}^{N} (upw)(x_k) \frac{\delta_{jk}^{(1)}}{h} + h \left(\frac{u(pw)'}{\phi'} \right)(x_j) \right| \quad (4.31)$$
$$\leq L_1 M^{1/2} \exp(-(\pi d\alpha M)^{1/2})$$

where L_1 is a constant depending on u, p, w, ϕ, and d.

(c) Let $u([S(j,h) \circ \phi]w)'' \in B(D)$ and B_{T_2} in (4.13) be zero. Then

$$\left| \int_a^b (u''[S(j,h) \circ \phi]w)(x)dx - h \sum_{k=-M}^{N} u(x_k) \left[\frac{\delta_{jk}^{(2)}}{h^2} (\phi'w)(x_k) \right. \right.$$
$$\left. \left. + \frac{\delta_{jk}^{(1)}}{h} \left(\frac{\phi''}{\phi'} w + 2w' \right)(x_k) \right] - h \left(\frac{w''u}{\phi'} \right)(x_j) \right| \quad (4.32)$$
$$\leq L_2 M \exp\left(-(\pi d\alpha M)^{1/2}\right)$$

where L_2 depends on u, w, ϕ, and d.

Proof

(a) From Theorem 4.2, for sufficiently small h, there exist constants R_p such that

$$C_p(h,d) \leq R_p h^{1-p}, \quad p = 0, 1, 2. \quad (4.33)$$

Substituting this (for $p = 0$) into the right-hand side of (4.24) and then using (4.29) leads directly to (4.30), where

$$L_0 \equiv \frac{R_0}{2} N(vw, D) \left(\frac{\pi d}{\alpha} \right)^{1/2}.$$

(b) Truncating (4.25) leads to

$$\left| \int_a^b (pu'[S(j,h) \circ \phi]w)(x)dx + h \sum_{k=-M}^{N} (upw)(x_k) \frac{\delta_{jk}^{(1)}}{h} \right.$$

$$+ \left. h\left(\frac{u(pw)'}{\phi'}\right)(x_j)\right| \le h \sum_{k=N+1}^{\infty} \left|(upw)(x_k)\frac{\delta_{jk}^{(1)}}{h}\right| \quad (4.34)$$

$$+ h \sum_{k=M+1}^{\infty} \left|(upw)(x_{-k})\frac{\delta_{j,-k}^{(1)}}{h}\right|$$

$$+ \frac{1}{2}\left[C_1(h,d)N(upw\phi',D) + C_0(h,d)N(u(pw)',D)\right]e^{-\pi d/h}.$$

The assumption in (4.27) (with $F = upw$), along with a computation similar to (2.39), leads to the bound

$$h \sum_{k=N+1}^{\infty} \left|(upw)(x_k)\frac{\delta_{jk}^{(1)}}{h}\right|$$

$$\le K \sum_{k=N+1}^{\infty} |\delta_{jk}^{(1)}| e^{-\beta kh}$$

$$\le K \sum_{k=N+1}^{\infty} e^{-\beta kh}$$

$$\le \frac{K}{\beta h} e^{-\beta Nh}$$

where $|\delta_{jk}^{(1)}| \le 1$ from (4.15). The second sum on the right-hand side of (4.34) is similarly bounded so that (4.34) takes the form

$$\left|\int_a^b (pu'[S(j,h)\circ \phi]w)(x)dx + h \sum_{k=-M}^{N} (upw)(x_k)\frac{\delta_{jk}^{(1)}}{h}\right.$$

$$+ \left. h\left(\frac{u(pw)'}{\phi'}\right)(x_j)\right| \le \frac{K}{h}\left[\frac{e^{-\beta Nh}}{\beta} + \frac{e^{-\alpha Mh}}{\alpha}\right]$$

$$+ \frac{1}{2}\left[C_1(h,d)N(upw\phi',D) + C_0(h,d)N(u(pw)',D)\right]e^{-\pi d/h}.$$

Using (4.33) and the selections (4.28) and (4.29) leads to (4.31), where

$$L_1 \equiv \frac{R_1}{2} N(upw\phi', D) + \left(\frac{\pi d}{\alpha}\right)^{1/2} \frac{R_0}{2} N(u(pw)', D)$$

$$+ \left(\frac{\alpha}{\pi d}\right)^{1/2} K \left(\frac{1}{\alpha} + \frac{1}{\beta}\right).$$

(c) A similar, but somewhat lengthier, computation (begin with (4.26)) uses the bound in (4.27) and $|\delta_{jk}^{(2)}| \leq \frac{\pi^2}{3}$ from (4.16) to arrive at (4.32), where

$$L_2 \equiv \frac{R_0}{2}\sqrt{\frac{\pi d}{\alpha}}\, N(uw'', D) + R_1 N(u[\phi''w + 2\phi'w'], D)$$
$$+ R_2 \sqrt{\frac{\alpha}{\pi d}}\, N(u(\phi')^2 w, D) + K\left(\frac{1}{\alpha} + \frac{1}{\beta}\right)\left(\frac{\alpha\pi}{3d} + \sqrt{\frac{\alpha}{\pi d}}\right).$$

∎

As in the last chapter, the choice for N in (4.28) is replaced by

$$N = \frac{\alpha}{\beta} M \qquad (4.35)$$

if $\alpha M/\beta$ is an integer. Recall the discussion following Theorem 2.14.

4.3 Discrete System Assembly and Error

The results in Theorem 4.4 contain all the approximations to formulate the discrete Sinc-Galerkin system for the problem (4.1). The Galerkin equations for $j = -M, \ldots, 0, \ldots, N$ are

$$\begin{aligned}
0 &= (Lu - f, S(j,h) \circ \phi) \\
&= -(u'', S(j,h) \circ \phi) + (pu', S(j,h) \circ \phi) \\
&\quad + (qu, S(j,h) \circ \phi) - (f, S(j,h) \circ \phi).
\end{aligned}$$

Approximation of the inner products by Theorem 4.4 yields the component form

$$\left| h \sum_{k=-M}^{N} u(x_k) \left[\frac{\delta_{jk}^{(2)}}{h^2} (\phi'w)(x_k) + \frac{\delta_{jk}^{(1)}}{h}\left(\frac{\phi''}{\phi'}w + 2w'\right)(x_k) \right] \right.$$
$$+ h\left(\frac{w''u}{\phi'}\right)(x_j) + h \sum_{k=-M}^{N} (upw)(x_k) \frac{\delta_{jk}^{(1)}}{h}$$
$$\left. + h\left(\frac{u(pw)'}{\phi'}\right)(x_j) - h\left(\frac{qwu}{\phi'}\right)(x_j) + h\left(\frac{fw}{\phi'}\right)(x_j) \right|$$

$$
\begin{aligned}
&\leq \left| -(u'', S_j) + \left\{ h \sum_{k=-M}^{N} u(x_k) \left[\frac{\delta_{jk}^{(2)}}{h^2} (\phi' w)(x_k) \right. \right. \right. \\
&\quad \left. \left. + \frac{\delta_{jk}^{(1)}}{h} \left(\frac{\phi''}{\phi'} w + 2w' \right)(x_k) \right] + h \left(\frac{w''u}{\phi'} \right)(x_j) \right\} \bigg| \\
&\quad + \left| (pu', S_j) + \left\{ h \sum_{k=-M}^{N} (upw)(x_k) \frac{\delta_{jk}^{(1)}}{h} + h \left(\frac{u(pw)'}{\phi'} \right)(x_j) \right\} \right| \\
&\quad + \left| (qu, S_j) - h \left(\frac{qwu}{\phi'} \right)(x_j) \right| + \left| (-f, S_j) + h \left(\frac{fw}{\phi'} \right)(x_j) \right| \\
&\leq \left(L_2 M + L_1 M^{1/2} + L_0 M^{-1/2} + L_0 M^{-1/2} \right) e^{-(\pi d \alpha M)^{1/2}} \\
&\leq (L_2 + L_1 + 2L_0) M e^{-(\pi d \alpha M)^{1/2}} \\
&\equiv C M e^{-(\pi d \alpha M)^{1/2}}.
\end{aligned}
$$

Deleting the error term of order $\mathcal{O}\left(M \exp(-(\pi d\alpha M)^{1/2}) \right)$, replacing $u(x_k)$ by u_k, and dividing by $(-h)$ yields the discrete sinc system

$$
\begin{aligned}
&\sum_{k=-M}^{N} \left[\frac{-1}{h^2} \delta_{jk}^{(2)} \phi'(x_k) w(x_k) \right. \\
&\quad \left. - \frac{1}{h} \delta_{jk}^{(1)} \left(\frac{\phi''(x_k) w(x_k)}{\phi'(x_k)} + 2w'(x_k) \right) \right] u_k \qquad (4.36) \\
&\quad - \frac{w''(x_j)}{\phi'(x_j)} u_j - \sum_{k=-M}^{N} \frac{1}{h} \delta_{jk}^{(1)} p(x_k) w(x_k) u_k \\
&\quad - \frac{(pw)'(x_j)}{\phi'(x_j)} u_j + \frac{q(x_j) w(x_j)}{\phi'(x_j)} u_j = \frac{f(x_j) w(x_j)}{\phi'(x_j)}.
\end{aligned}
$$

This system is identical to that generated from orthogonalizing the residual via $(Lu_m - f, S(j,h) \circ \phi) = 0$. The system in (4.36) is more conveniently recorded by defining the vectors

$$\vec{u} \equiv (u_{-M}, u_{-M+1}, \ldots, u_0, \ldots, u_N)^T$$

and

$$\vec{f} \equiv (f(x_{-M}), f(x_{-M+1}), \ldots, f(x_0), \ldots, f(x_N))^T.$$

Then define the $m \times m$ $(m = M + N + 1)$ Toeplitz matrices (constant diagonals in Definition A.15)

$$I^{(\ell)} \equiv \left[\delta_{jk}^{(\ell)} \right], \quad \ell = 0, 1, 2$$

THE SINC-GALERKIN METHOD

and the diagonal matrix

$$D(g) \equiv \begin{bmatrix} g(x_{-M}) & & & & \\ & \ddots & & & \\ & & g(x_0) & & \\ & & & \ddots & \\ & & & & g(x_N) \end{bmatrix}.$$

The matrix $I^{(0)}$ is the identity matrix. The matrices $I^{(1)}$ and $I^{(2)}$ take the form (use (4.15) and (4.16))

$$I^{(1)} = \begin{bmatrix} 0 & -1 & \frac{1}{2} & \cdots & \frac{(-1)^{m-1}}{m-1} \\ 1 & & & & \vdots \\ -\frac{1}{2} & & \ddots & & \frac{1}{2} \\ \vdots & & & & -1 \\ \frac{(-1)^m}{m-1} & \cdots & -\frac{1}{2} & 1 & 0 \end{bmatrix} \qquad (4.37)$$

$$I^{(2)} = \begin{bmatrix} -\frac{\pi^2}{3} & 2 & \frac{-2}{2^2} & \cdots & \frac{-2(-1)^{m-1}}{(m-1)^2} \\ 2 & & & & \vdots \\ \frac{-2}{2^2} & & \ddots & & \frac{-2}{2^2} \\ \vdots & & & & 2 \\ \frac{-2(-1)^{m-1}}{(m-1)^2} & \cdots & \frac{-2}{2^2} & 2 & -\frac{\pi^2}{3} \end{bmatrix}. \qquad (4.38)$$

The properties of these matrices are given in Theorems 4.18 and 4.19.
The system in (4.36) for the problem (4.1) given by

$$Lu(x) \equiv -u''(x) + p(x)u'(x) + q(x)u(x) = f(x), \quad a < x < b$$
$$u(a) = u(b) = 0$$

takes the form

$$\left\{\frac{-1}{h^2} I^{(2)} D(\phi'w) - \frac{1}{h} I^{(1)} D\left(\frac{\phi''w}{\phi'} + 2w' + pw\right)\right.$$

$$\left. - D\left(\frac{w'' + (pw)' - qw}{\phi'}\right)\right\} \vec{u} = D\left(\frac{w}{\phi'}\right) \vec{f}.$$
(4.39)

For reasons that will emerge in the development, it is helpful to single out the portion of the coefficient matrix in this system that corresponds to the second derivative. This is defined by

$$\mathcal{A}(w) \equiv \frac{-1}{h^2} I^{(2)} - \frac{1}{h} I^{(1)} D\left(\frac{\phi''}{(\phi')^2} + \frac{2w'}{\phi'w}\right) - D\left(\frac{w''}{(\phi')^2 w}\right). \quad (4.40)$$

If the assumed approximate solution of (4.1) is

$$u_m(x) = \sum_{k=-M}^{N} u_k S(k,h) \circ \phi(x), \quad m = M + N + 1 \quad (4.41)$$

then the discrete Sinc-Galerkin system for the determination of the $\{u_k\}_{k=-M}^{N}$ is (4.39). The case thoroughly analyzed in [12] selects for a weight in (4.39) the function

$$w(x) = \frac{1}{\phi'(x)}. \quad (4.42)$$

In this case, (4.39) takes the form

$$\left\{\mathcal{A}\left(\frac{1}{\phi'}\right) - \frac{1}{h} I^{(1)} D\left(\frac{p}{\phi'}\right) + D\left(\frac{-1}{\phi'}\left(\frac{p}{\phi'}\right)' + \frac{q}{(\phi')^2}\right)\right\} \vec{u}$$

$$= D\left(\frac{1}{(\phi')^2}\right) \vec{f} \quad (4.43)$$

where, from (4.40),

$$\mathcal{A}\left(\frac{1}{\phi'}\right) \equiv \frac{-1}{h^2} I^{(2)} + \frac{1}{h} I^{(1)} D\left(\frac{\phi''}{(\phi')^2}\right) + D\left(\frac{-1}{(\phi')}\left(\frac{1}{\phi'}\right)''\right) \quad (4.44)$$

since $\phi''/(\phi')^2 + 2(1/(\phi'))' = -\phi''/(\phi')^2$.

To see why such a weight function is selected, consider the problem (4.1) on $(0,1)$. Assume this problem has regular singular points at $x = 0$ and $x = 1$ (see Exercise 4.1 for the definition). One of the assumptions in the derivation of (4.43) is that the boundary terms B_{T_j} ($j = 1, 2$) in (4.11) and (4.13) vanish. If the map $\phi(x) = \ell n(x/(1-x))$ is used in the evaluation of these boundary terms, then Exercise 4.1 shows that each of the boundary terms is zero in the case that the problem has (at worst) regular singular points.

The inequality (4.27) is satisfied if there are positive constants α, β, and $\hat{K}$ such that

$$|p(x)u(x)| \leq \hat{K} \begin{cases} x^{\alpha-1}, & x \in (0, 1/2) \\ (1-x)^{\beta-1}, & x \in [1/2, 1) \end{cases}$$

which, in terms of the solution u, reads

$$|u(x)| \leq \hat{K} \begin{cases} x^{\alpha}, & x \in (0, 1/2) \\ (1-x)^{\beta}, & x \in [1/2, 1) \end{cases} \quad (4.45)$$

if the problem has regular singular points. That is, if (4.45) is satisfied, then (4.27) is satisfied.

The system (4.43) may be easily solved by a variety of methods. In the results that follow in this chapter direct elimination methods are used. Iterative methods, coupled with multigrid techniques, were studied in [8].

In Table 4.1 below, the relevant quantities to fill the matrices in (4.43) for the various maps ϕ of Table 3.7 are summarized. In the examples below the maximum errors over the set of sinc gridpoints

$$\mathcal{S} = \{x_{-M}, x_{-M+1}, \ldots, x_N\};$$

$$x_k = \frac{e^{kh}}{e^{kh}+1}, \quad k = -M, \ldots, N$$

are reported as

$$\|E_{\mathcal{S}}(h)\| = \max_{-M \leq k \leq N} |u(x_k) - u_k|. \quad (4.46)$$

In the first three examples (which are on the interval $(0,1)$) the errors are also reported on a uniform grid

$$U = \{z_0, z_1, \ldots, z_{100}\};$$

$$z_k = kh_u, \quad h_u = .01, \quad k = 0, 1, \ldots, 100.$$

Differential Equation:

$$Lu(x) \equiv -u''(x) + p(x)u'(x) + q(x)u(x) = f(x), \quad a < x < b$$
$$u(a) = u(b) = 0$$

Sinc-Galerkin system for $w = 1/\phi'$:

$$\mathcal{A}\left(\frac{1}{\phi'}\right) \equiv \frac{-1}{h^2}I^{(2)} + \frac{1}{h}I^{(1)}D\left(\frac{\phi''}{(\phi')^2}\right) + D\left(\frac{-1}{\phi'}\left(\frac{1}{\phi'}\right)''\right)$$

$$\left\{\mathcal{A}\left(\frac{1}{\phi'}\right) - \frac{1}{h}I^{(1)}D\left(\frac{p}{\phi'}\right) + D\left(\frac{-1}{\phi'}\left(\frac{p}{\phi'}\right)' + \frac{q}{(\phi')^2}\right)\right\}\vec{u}$$
$$= D\left(\frac{1}{(\phi')^2}\right)\vec{f}$$

a	b	ϕ	$\frac{\phi''}{(\phi')^2}$	$\frac{1}{\phi'}$	$-\frac{1}{\phi'}\left(\frac{1}{\phi'}\right)''$
a	b	$\ln\left(\frac{x-a}{b-x}\right)$	$\frac{2x-a-b}{b-a}$	$\frac{(x-a)(b-x)}{b-a}$	$\frac{2(x-a)(b-x)}{(b-a)^2}$
0	1	$\ln\left(\frac{x}{1-x}\right)$	$2x-1$	$x(1-x)$	$2x(1-x)$
0	∞	$\ln(x)$	-1	x	0
0	∞	$\ln(\sinh(x))$	$-\text{sech}^2(x)$	$\tanh(x)$	$\frac{2\tanh^2(x)}{\cosh^2(x)}$
$-\infty$	∞	x	0	1	0
$-\infty$	∞	$\sinh^{-1}(x)$	$\frac{-x}{(x^2+1)^{1/2}}$	$(1+x^2)^{1/2}$	$\frac{-1}{x^2+1}$

Table 4.1 Components for the discrete system in (4.43).

These errors are given as

$$\|E_U(h_u)\| = \max_{0 \le k \le 100} |u(z_k) - u_m(z_k)|. \qquad (4.47)$$

Recalling the discussion in Example 3.6, these three examples again illustrate that the error $\|E_S(h)\|$ dominates the error $\|E_U(h_u)\|$. Using the interpolatory property of the sinc approximation, $u_m(x_k) = u_k$, again shows that the former is computationally more accessible than is the latter.

Example 4.5 The discretization for the problem on $(0,1)$ given by

$$-u''(x) - \frac{1}{6x} u'(x) + \frac{1}{x^2} u(x) = \frac{19}{6}\sqrt{x}$$

$$u(0) = u(1) = 0 \qquad (4.48)$$

which has solution $u(x) = x^{3/2}(1-x)$ takes the form (from Table 4.1)

$$\left\{ \frac{-1}{h^2} I^{(2)} + \frac{1}{h} I^{(1)} D\left(\frac{11x - 5}{6}\right) - D\left((1-x)(1 + \frac{5}{6}x)\right) \right\} \vec{u}$$

$$= D\left(\frac{19}{6} x^{5/2}(1-x)^2\right) \vec{1}$$

where

$$x_k = e^{kh}/(1 + e^{kh}).$$

The solution $\vec{u} = (u_{-M}, \ldots, u_0, \ldots, u_N)^T$ gives the coefficients in the approximate Sinc-Galerkin solution

$$u_m(x) = \sum_{k=-M}^{N} u_k S(k,h) \circ \left(\ell n\left(\frac{x}{1-x}\right)\right).$$

This problem has a regular singular point at $x = 0$ and the quantities α and β are determined from (4.45). Since the true solution of (4.48) is $u(x) = x^{3/2}(1-x)$ one can select $\alpha = 3/2$ and $\beta = 1$. The discussion in Section 4.5 gives selections of α and β based on series solutions of (4.1). Recalling (4.35) and (4.29), the selections $N = 3M/2$ and $h = \pi/\sqrt{3M}$ ($d = \pi/2$) are made in Table 4.2. Notice that the error on the sinc grid $\|E_S(h)\|$ is a bit larger than the error on the uniform grid $\|E_U(h_u)\|$. Not only in this example, but in many others that have been tested, this seems always to be the case. That is, the error on the sinc grid very faithfully represents, and is usually a little larger than, the numerical uniform error. This, as pointed out in Example 3.6, is a function of the uniform grid size and the nature of the solution u. As a rule of thumb, one typically has a good estimate of the uniform error by just checking the error on the sinc grid. Assigning the value one to each of α and β and computing with the mesh $h = \pi/\sqrt{2M}$ yields results with an error that is about a digit worse than the results listed in the column labeled $\|E_S(h)\|$ in Table 4.2. This will be illustrated in Example 4.16 where a reason for the choice $\alpha = \beta = 1$ is given.

M	N	$h = \pi/\sqrt{3M}$	$\|E_S(h)\|$	$\|E_U(h_u)\|$
4	6	0.907	$.208 \times 10^{-2}$	$.200 \times 10^{-2}$
8	12	0.641	$.262 \times 10^{-3}$	$.254 \times 10^{-3}$
16	24	0.453	$.127 \times 10^{-4}$	$.124 \times 10^{-4}$
32	48	0.321	$.157 \times 10^{-6}$	$.154 \times 10^{-6}$

Table 4.2 Errors in the Sinc-Galerkin solution of (4.48) with weight function $w(x) = 1/\phi'(x)$.

∎

Example 4.6 The solution u of the convection problem

$$-u''(x) + \kappa u'(x) = f(x)$$
$$u(0) = u(1) = 0$$
(4.49)

is, for large values of κ, difficult to compute. This may be forecast by an inspection of the exact solution corresponding to $f(x) = \kappa$,

$$u(x) = x - \frac{e^{\kappa x} - 1}{e^{\kappa} - 1}.$$

THE SINC-GALERKIN METHOD 121

The solution is very slowly changing over most of the interval $(0,1)$ but abruptly turns at $x_\kappa = (1/\kappa)\ell n((e^\kappa - 1)/\kappa)$ ($x_{10} = .77$ and $x_{100} = .95$) to return to the zero boundary value at $x = 1$.

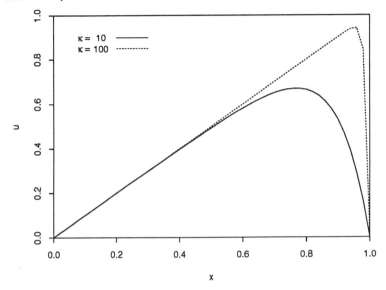

Figure 4.1 Computed solution of (4.49) with $M = N = 8$ and $\kappa = 10$, 100.

Again, from Table 4.1, the Sinc-Galerkin discretization of the problem reads

$$\left\{\frac{-1}{h^2}I^{(2)} + \frac{1}{h}I^{(1)}D\left(2x - 1 - \kappa x(1-x)\right)\right.$$
$$\left. + D(2x(1-x) + \kappa x(1-x)(2x-1))\right\}\vec{u}$$
$$= D\left(\kappa(x(1-x))^2\right)\vec{1}.$$

The solution $\vec{u} = (u_{-M}, \ldots, u_0, \ldots, u_N)^T$ gives the coefficients in the approximate Sinc-Galerkin solution

$$u_m(x) = \sum_{k=-M}^{N} u_k S(k,h) \circ \left(\ell n\left(\frac{x}{1-x}\right)\right).$$

Using $\alpha = \beta = 1$ in (4.30) and (4.36), respectively, gives the mesh selection $h = \pi/\sqrt{2M}$ and $N = M$. Figure 4.1 has a graph of the computed solution u_m, where $m = 17$, corresponding to the values $\kappa = 10$ and 100. The graphs are indistinguishable from the true solution. In Table 4.3 the errors are reported on the sinc grid $\|E_S(h)\|$ and on the uniform grid $\|E_U(h_u)\|$ for $\kappa = 10$ and $\kappa = 100$.

		$\kappa = 10$	
M	$h = \pi/\sqrt{2M}$	$\|E_S(h)\|$	$\|E_U(h_u)\|$
4	1.111	$.352 \times 10^{-1}$	$.356 \times 10^{-1}$
8	0.785	$.826 \times 10^{-2}$	$.761 \times 10^{-2}$
16	0.555	$.772 \times 10^{-3}$	$.700 \times 10^{-3}$
32	0.393	$.223 \times 10^{-4}$	$.429 \times 10^{-5}$

		$\kappa = 100$	
M	$h = \pi/\sqrt{2M}$	$\|E_S(h)\|$	$\|E_U(h_u)\|$
4	1.111	$.150 \times 10^{-0}$	$.186 \times 10^{-0}$
8	0.785	$.829 \times 10^{-1}$	$.359 \times 10^{-1}$
16	0.555	$.840 \times 10^{-2}$	$.318 \times 10^{-2}$
32	0.393	$.246 \times 10^{-3}$	$.914 \times 10^{-4}$

Table 4.3 Errors in the Sinc-Galerkin solution corresponding to $\kappa = 10$ and 100 in (4.49).

■

The status of the Sinc-Galerkin method with the weight function $w(x) = 1/\phi'(x)$ is thoroughly reviewed in [13]. The excellent convergence in the previous two examples is guaranteed by the following theorem [12].

Theorem 4.7 *Assume that the coefficients p, q, and f in*

$$Lu(x) \equiv -u''(x) + p(x)u'(x) + q(x)u(x) = f(x), \quad a < x < b$$
$$u(a) = u(b) = 0$$

and the unique solution u are analytic in the simply connected domain D. Let ϕ be a conformal one-to-one map of D onto D_S. Assume also that $f/\phi' \in B(D)$, and $uF \in B(D)$ where

$$F = (1/\phi')'', \; (\phi''/\phi'), \; \phi', \; (p/\phi')', \; p, \; (q/\phi'). \qquad (4.50)$$

Suppose there are positive constants C, α, and β so that

$$|u(x)| \le C \begin{cases} \exp(-\alpha|\phi(x)|), & x \in \Gamma_a \\ \exp(-\beta|\phi(x)|), & x \in \Gamma_b. \end{cases} \qquad (4.51)$$

Define the Sinc-Galerkin solution by

$$u_m(x) = \sum_{k=-M}^{N} u_k S(k,h) \circ \phi(x)$$

where

$$N = \left[\left|\frac{\alpha}{\beta} M + 1\right|\right], \quad h = \left(\frac{\pi d}{\alpha M}\right)^{1/2}$$

and the $\{u_k\}_{k=-M}^{N}$ are found from the discrete system in Table 4.1. Then

$$\|u - u_m\|_\infty \le KM^2 \exp(-(\pi d \alpha M)^{1/2}). \qquad (4.52)$$

∎

The assumptions in (4.50) appear a bit elaborate. Part of the conclusion of Exercise 4.2 is that if $u \in B(D)$, then the first three functions in (4.50) (times u) are in $B(D)$; the last three (times u), as well as f/ϕ', are, as long as

$$f(x), p(x) = \mathcal{O}\left(\frac{1}{x-a}\right), \quad x \to a^+ \qquad (4.53)$$

and

$$q(x) = \mathcal{O}\left(\frac{1}{(x-a)^2}\right), \quad x \to a^+ \qquad (4.54)$$

with a similar statement as $x \to b^-$ (if b is finite). That is, if the differential equation has at worst a regular singular point at a (and/or b), then $f/\phi' \in B(D)$ and (4.50) holds. In the case of $b = +\infty$, there are some minimal integrability requirements on the functions in the proof of Theorem 4.3 that are sorted out in Exercise 4.3 for the map $ln(x)$. The transcription of the condition on the right-hand side of (4.51) for the maps $\phi(x)$, as well as the identification of the intervals Γ_a and Γ_b, are recorded in Table 3.7.

Example 4.8 A differential equation whose solution exhibits a logarithmic singularity is provided by

$$-u''(x) + \frac{1}{x^2} u(x) = \frac{ln(x) - 1}{x}, \quad 0 < x < 1 \tag{4.55}$$

$$u(0) = u(1) = 0.$$

That is, the derivative of the solution $u(x) = x\, ln(x)$, diverges like $ln(x)$ as $x \to 0^+$. The discrete system, taken from Table 4.1, reads

$$\left\{ \frac{-1}{h^2} I^{(2)} + \frac{1}{h} I^{(1)} D(2x - 1) + D(1 - x^2) \right\} \vec{u} \tag{4.56}$$

$$= D\left(x(1-x)^2 (ln(x) - 1) \right) \vec{1}.$$

To determine the quantities α and β, one needs to check

$$|u(x)| \le \hat{K} \begin{cases} x^\alpha, & x \in (0, 1/2) \\ (1-x)^\beta, & x \in [1/2, 1). \end{cases}$$

Since the true solution satisfies

$$x\, ln(x) \le 1 - x$$

and

$$x\, ln(x) \le x^{1-\varepsilon}, \quad 1 - \varepsilon > 0$$

for x near one and zero, respectively, one can select $\beta = 1$ and $\alpha = 1 - \varepsilon$. Table 4.4 below uses $\alpha = 1$. From (4.35) and (4.29) the selections $N = M$ and $h = \pi/\sqrt{2M}$ ($d = \pi/2$) are made. Note that the error in Table 4.4 is about the same as the error in Table 4.3

THE SINC-GALERKIN METHOD   125

($\kappa = 10$) in spite of the logarithmic singularity. The solution in the previous example is analytic at the origin.

$M = N$	$h = \pi/\sqrt{2M}$	$\|E_S(h)\|$	$\|E_U(h_u)\|$
4	1.111	$.149 \times 10^{-1}$	$.168 \times 10^{-1}$
8	0.785	$.451 \times 10^{-2}$	$.154 \times 10^{-2}$
16	0.555	$.608 \times 10^{-3}$	$.833 \times 10^{-4}$
32	0.393	$.262 \times 10^{-4}$	$.244 \times 10^{-5}$

Table 4.4 Errors in the Sinc-Galerkin solution of (4.55).

∎

Example 4.9 Consider the differential equation

$$-u''(x) + u(x) = e^{-x}(10\cos(5x) + 25\sin(5x)), \quad 0 < x < \infty \quad (4.57)$$

$$u(0) = \lim_{x \to \infty} u(x) = 0$$

which has solution $u(x) = e^{-x}\sin(5x)$. The discrete system using the map $\phi(x) = \ell n(\sinh(x))$ is

$$\left\{ \frac{-1}{h^2} I^{(2)} + \frac{1}{h} I^{(1)} D\left(\frac{-1}{\cosh^2(x)}\right) \right.$$
$$\left. + D\left(\tanh^2(x)\left[\frac{2}{\cosh^2(x)} + 1\right]\right) \right\} \vec{u}$$
$$= D\left(\tanh^2(x)\right) \vec{f}.$$

From Table 3.7 the parameter choices are $\alpha = \beta = 1$, which lead to $N = M$ and $h = \pi/\sqrt{2M}$ ($d = \pi/2$). The final column of Table 4.5 records NOC, the ratio of $\|E_S(h)\|$ to $CR = \mathcal{O}(M^2 \exp(-\pi\sqrt{M}))$. Recall from Examples 3.6 and 3.18 that the convergence rate there, as well as here, gives an accurate estimate of the error. The discrete system based on the map $\phi(x) = \ell n(x)$ does a much poorer job on this example (Exercise 4.10).

$M=N$	$h=\pi/\sqrt{2M}$	$\|E_S(h)\|$	CR	NOC
4	1.111	$.154\times10^{+1}$	$.299\times10^{-1}$	51.5
8	0.785	$.198\times10^{+0}$	$.885\times10^{-2}$	22.4
16	0.555	$.596\times10^{-2}$	$.892\times10^{-3}$	6.68
32	0.393	$.100\times10^{-3}$	$.196\times10^{-4}$	5.10

Table 4.5 Results of the implementation of the $ln(\sinh(x))$ map for the oscillatory solution of (4.57).

■

Whereas a better approximation was obtained in the previous example using the map $\phi(x) = ln(\sinh(x))$ (compared to the map $\phi(x) = ln(x)$), a complementary scenario is provided by differential equations with a rational solution.

Example 4.10 The discretization for the problem on $(0,\infty)$ given by

$$-u''(x) + \frac{x}{x^2+1}u'(x) + \frac{1}{x^2+1}u(x) = \frac{-2x(x^2-4)}{(x^2+1)^3} \quad (4.58)$$

$$u(0) = \lim_{x\to\infty} u(x) = 0$$

(which has solution $u(x) = \frac{x}{x^2+1}$) takes the form (from Table 4.1 with map $\phi(x) = ln(x)$)

$$\left\{\frac{-1}{h^2}I^{(2)} + \frac{1}{h}I^{(1)}D\left(\frac{-(2x^2+1)}{x^2+1}\right) + D\left(\frac{x^2(x^2-1)}{(x^2+1)^2}\right)\right\}\vec{u}$$

$$= D\left(\frac{-2x^3(x^2-4)}{(x^2+1)^3}\right)\vec{1}$$

where

$$x_k = e^{kh}.$$

THE SINC-GALERKIN METHOD

The solution $\vec{u} = (u_{-M}, \ldots, u_0, \ldots, u_N)^T$ gives the coefficients in the approximate Sinc-Galerkin solution

$$u_m(x) = \sum_{k=-M}^{N} u_k S(k, h) \circ (\ell n(x)).$$

Notice that if one uses the system in Table 4.1 corresponding to the $\phi(x) = \ell n(\sinh(x))$, then one will not obtain the exponential convergence of the method due to the rational character of the solution. Recall the assumed growth condition in Table 3.7 for the $\phi(x) = \ell n(\sinh(x))$ requires exponential decay of the solution to guarantee an exponential decay of the convergence rate.

The quantities α and β are determined from (4.45). Since the true solution of (4.58) is $u(x) = \frac{x}{x^2+1}$, one can select $\alpha = 1$ and $\beta = 1$. Recalling (4.35) and (4.29), the selections $N = M$ and $h = \pi/\sqrt{2M}$ ($d = \pi/2$) are made in Table 4.6.

$M = N$	$h = \pi/\sqrt{2M}$	$\|E_\mathcal{S}(h)\|$
4	1.111	$.249 \times 10^{-1}$
8	0.785	$.349 \times 10^{-2}$
16	0.555	$.240 \times 10^{-3}$
32	0.393	$.547 \times 10^{-5}$

Table 4.6 Calculation of the rational solution of (4.58) using the map $\phi(x) = \ell n(x)$.

■

Notice that the boundary value problem in Example 4.8 is self-adjoint but that the discrete system (4.39) generating the approximate solution is not symmetric due to the presence of the matrix $I^{(1)}$. While a symmetric approximating system is not necessary for a good approximation, it is computationally efficient and analytically advantageous for establishing error bounds.

If in (4.1) $p(x) \equiv 0$ on (a,b), the boundary value problem takes the form

$$L_s u(x) \equiv -u''(x) + q(x)u(x) = f(x), \quad a < x < b \tag{4.59}$$
$$u(a) = u(b) = 0.$$

This is a self-adjoint problem, but the discrete Sinc-Galerkin system (4.43) is never symmetric (for all the maps considered here $\phi''/(\phi')^2$ is never identically zero). If instead of the weight function $w(x) = (\phi'(x))^{-1}$ one selects [4]

$$w(x) = \frac{1}{\sqrt{\phi'(x)}} \tag{4.60}$$

in the system (4.39), the result is

$$\left[\mathcal{A}\left(\frac{1}{\sqrt{\phi'}}\right) + D\left(\frac{q}{(\phi')^2}\right) \right] \vec{y} = D\left(\frac{1}{(\phi')^{3/2}}\right) \vec{f} \tag{4.61}$$

where

$$\mathcal{A}\left(\frac{1}{\sqrt{\phi'}}\right) = \frac{-1}{h^2} I^{(2)} + D\left(\left(\frac{-1}{(\phi')^{3/2}}\right)\left(\frac{1}{\sqrt{\phi'}}\right)''\right) \tag{4.62}$$

and vector $\vec{y}$ is given by

$$\vec{y} \equiv D\left(\sqrt{\phi'}\right) \vec{u}^s. \tag{4.63}$$

Since $I^{(2)}$ is symmetric, the Sinc-Galerkin system in (4.61) is a symmetric discrete system for the determination of the coefficients in

$$u_m^s(x) \equiv \sum_{k=-M}^{N} u_k^s S(k,h) \circ \phi(x). \tag{4.64}$$

Instead of a development for the error in approximating the true solution by (4.64) as was done earlier, it is more direct to relate the present situation to the previous development using (4.63). Make the change of dependent variable

$$y(x) = (\phi'(x))^{1/2} u(x)$$

in (4.59) to find

$$L_s y(x) \equiv -y''(x) - 2\sqrt{\phi'(x)}\left(\frac{1}{\sqrt{\phi'(x)}}\right)' y'(x)$$
$$- \left(\sqrt{\phi'(x)}\left(\frac{1}{\sqrt{\phi'(x)}}\right)'' - q(x)\right) y(x) \quad (4.65)$$
$$= \sqrt{\phi'(x)} f(x),$$
$$y(a) = y(b) = 0.$$

The application of Theorem 4.7 to (4.65) gives the following theorem from [4].

Theorem 4.11 *Assume that the coefficients $\sqrt{\phi'}\left(1/\sqrt{\phi'}\right)'$, $\sqrt{\phi'} f$, and $\left(\sqrt{\phi'}\left(1/\sqrt{\phi'}\right)'' - q\right)$ in (4.65) and the unique solution y are analytic in the simply connected domain D. Let ϕ be a conformal one-to-one map of D onto D_S. Assume also that $f/\sqrt{\phi'} \in B(D)$ and $yF \in B(D)$, where*

$$F = (1/\phi')'',\ \phi',\ \phi''/\phi',\ 1/\sqrt{\phi'}\left(1/\sqrt{\phi'}\right)'',\ q/\phi'.$$

Suppose there are positive constants C_s, α_s, and β_s so that

$$|y(x)| = |u(x)(\phi'(x))^{1/2}| \leq C_s \begin{cases} \exp(-\alpha_s |\phi(x)|), & x \in \Gamma_a \\ \exp(-\beta_s |\phi(x)|), & x \in \Gamma_b. \end{cases} \quad (4.66)$$

If

$$y_m(x) = \sum_{k=-M}^{N} y_k S(k,h) \circ \phi(x) \quad (4.67)$$

is the assumed Sinc-Galerkin solution and the $\{y_k\}_{k=-M}^{N}$ are determined from the discrete system in (4.61) with

$$N = \left[\left|\frac{\alpha_s}{\beta_s} M + 1\right|\right],\quad h_s = \left(\frac{\pi d}{\alpha_s M}\right)^{1/2} \quad (4.68)$$

then

$$\|y - y_m\|_\infty \leq K_s M^2 \exp\left(-(\pi d \alpha_s M)^{1/2}\right). \quad (4.69)$$

Proof Apply the Sinc-Galerkin method to (4.65) using the weight $w(x) = (\phi'(x))^{-1}$. This gives the linear system (4.61). Thus Theorem 4.7 applies to $y(x) = \sqrt{\phi'(x)}\,u(x)$.

■

Note that since $y(x) = \sqrt{\phi'(x)}\,u(x)$, $(\phi'(x))^{-1/2} \leq 1/2$ for all $x \in [0,1]$, (4.69) yields

$$\|u - u_m^s\|_\infty \leq L_s M^2 \exp\left(-(\pi d\alpha_s M)^{1/2}\right).$$

The identity $y_k = \sqrt{\phi'(x_k)}\,u_k^s$ from (4.63) shows that at the nodes x_k

$$|u(x_k) - u_m^s(x_k)| = \frac{1}{\sqrt{\phi'(x_k)}}\,|y(x_k) - y_m(x_k)|$$

$$\leq \frac{K_s M^2}{\sqrt{\phi'(x_k)}}\,e^{-(\pi d\alpha_s M)^{1/2}}.$$

In the special case $\phi(x) = \ell n(x/(1-x))$, it is shown in [4] that

$$\frac{1}{\sqrt{\phi'(x_k)}}\,e^{-(\pi d\alpha_s M)^{1/2}} = \mathcal{O}\left(e^{-(\pi d\alpha M)^{1/2}}\right).$$

Hence, at the nodes, the accuracy in the computed solution u_m^s in (4.64) is (asymptotically) the same as the accuracy in the computed solution u_m given in Theorem 4.7. Thus the selection $h = (\pi d/(\alpha M))^{1/2}$ (instead of h_s in (4.68)) should lead to errors for u_m^s that are effectively the same as errors for u_m. Numerical results substantiate this remark. Hence, the selection of the weight $w(x) = (\phi'(x))^{-1/2}$ for the problem (4.59) is recommended. The entries for the discrete system (4.61) are displayed in Table 4.7. It should be pointed out that the selection of the weight in (4.60) does not lead to a symmetric discrete system for the more general boundary value problem (4.1). Make the change of variable

$$u(x) = \sigma(x)\exp\left(-\frac{1}{2}\int^x p(t)dt\right)$$

in (4.1). Then a new boundary value problem in the dependent variable σ arises. It has the form

Differential Equation:

$$L_s u(x) \equiv -u''(x) + q(x)u(x) = f(x), \quad a < x < b$$
$$u(a) = u(b) = 0$$

Sinc-Galerkin System for $w = 1/\sqrt{\phi'}$:

$$A\left(\frac{1}{\sqrt{\phi'}}\right) = \frac{-1}{h^2} I^{(2)} + D\left(\left(\frac{-1}{(\phi')^{3/2}}\right)\left(\frac{1}{\sqrt{\phi'}}\right)''\right)$$

$$\left[A\left(\frac{1}{\sqrt{\phi'}}\right) + D\left(\frac{q}{(\phi')^2}\right)\right]\vec{y} = D\left(\frac{1}{(\phi')^{3/2}}\right)\vec{f}$$

a	b	ϕ	$\frac{1}{\phi'}$	$\frac{-1}{(\phi')^{3/2}}\left(\frac{1}{\sqrt{\phi'}}\right)''$
a	b	$\ln\left(\frac{x-a}{b-x}\right)$	$\frac{(x-a)(b-x)}{b-a}$	$1/4$
0	1	$\ln\left(\frac{x}{1-x}\right)$	$x(1-x)$	$1/4$
0	∞	$\ln(x)$	x	$1/4$
0	∞	$\ln(\sinh(x))$	$\tanh(x)$	$\frac{4\cosh^2(x)-3}{4\cosh^4(x)}$
$-\infty$	∞	x	1	0
$-\infty$	∞	$\sinh^{-1}(x)$	$(1+x^2)^{1/2}$	$\frac{x^2-2}{4(x^2+1)}$

Table 4.7 Components for the symmetric discrete system in (4.61).

$$-\sigma''(x) + g(x)\sigma(x) = \mu(x)$$
$$\sigma(a) = \sigma(b) = 0$$

where
$$g(x) = \frac{-3}{4}(p(x))^2 + \frac{p'(x)}{2} + q(x)$$

and
$$\mu(x) = f(x)\exp\left(\frac{1}{2}\int^x p(t)dt\right). \quad (4.70)$$

It may be argued that using the weight function $w(x) = [\phi'(x)]^{-1/2}$ for this transformed problem leads to a symmetric sinc system for the approximation of σ. While this point of view may have merit, there are a few reasons why a direct application of the procedure in Theorem 4.7 seems the preferable approach. The gain of a symmetric system comes at the expense of the numerical evaluation of the $M + N + 1$ indefinite integrals in (4.70) to fill the right-hand side of the discrete system. More important than a little extra computation is the reality that it is not always a good idea to strive for a symmetric approximating system when the problem to be solved is not inherently symmetric. As a specific case in point, the boundary value problem in Example 4.6 can be converted to a self-adjoint problem where the integrals in (4.70) can be computed in closed form. Application of the symmetric sinc system to this transformed problem leads to poor results for large κ due to the conditioning of the coefficient matrix in (4.61). As a final point, the system in (4.43) does an awfully good job on the general problem (4.1), and it does not seem to be a good idea to attempt to fix something that is not broken.

Using the bounds in (4.66) and Table 3.7 the connection between α_s (β_s) and α (β) in (4.51) is

$$\alpha_s = \alpha - 1/2$$
$$\beta_s = \beta - 1/2. \quad (4.71)$$

Since α_s and β_s are assumed positive, the Sinc-Galerkin method based on the weight function $w(x) = (\phi'(x))^{-1}$ applies to a wider class of problems.

THE SINC-GALERKIN METHOD

Example 4.12 The self-adjoint boundary value problem in Example 4.8 is amenable to approximate solution via the technique described in Theorem 4.11. Table 4.7 includes all the necessary components to fill the discrete system for the problem

$$-u''(x) + \frac{1}{x^2} u(x) = \frac{\ell n(x) - 1}{x}, \quad 0 < x < 1$$
$$u(0) = u(1) = 0.$$

The matrix system

$$\left\{ \frac{-1}{h^2} I^{(2)} + D(1/4 + (1-x)^2) \right\} D\left((x(1-x))^{-1/2}\right) \vec{u}^{\,s}$$
$$= D\left(\sqrt{x}\,(1-x)^{3/2}(\ell n(x) - 1)\right) \vec{1}$$

for the coefficients $\vec{u}^{\,s}$ in

$$u_m^s(x) = \sum_{k=-M}^{N} u_k^s S(k,h) \circ \left(\ell n\left(\frac{x}{1-x}\right)\right)$$

is symmetric, as described above.

Since, from Example 4.8, $\alpha = \beta = 1$, it follows from (4.71) that $\alpha_s = \beta_s = 1/2$. Hence, from (4.35) $N = M$ and using (4.68) one selects $h_s = \pi/\sqrt{M}$ ($d = \pi/2$). In Table 4.8 the errors listed under $\|E_S(h)\|$ and $\|E_U(h_u)\|$ are repeated from Table 4.4. Based on the discussion following Theorem 4.11, listed under $\|E_S^s(h)\|$ are the errors

$$\|E_S^s(h)\| = \max_{-M \leq k \leq N} |u(x_k) - u_k^s|$$

with $h = (\pi d/(\alpha M))^{1/2} = \pi/\sqrt{2M}$ (the selection of h used in Table 4.4). The two columns labeled $\|E_S^s(h_s)\|$ and $\|E_U^s(h_u)\|$ are the analogues of $\|E_S(h)\|$ and $\|E_U(h_u)\|$ with $h_s = (\pi d/(\alpha_s M))^{1/2} = \pi/\sqrt{M}$. The typical values of $M = 4, 8, 16$, and 32 are used. In Table 4.8 the temporary notation $.ddd - q = .ddd \times 10^{-q}$ has been introduced to shorten the display.

$$h = \left(\frac{\pi d}{\alpha M}\right)^{1/2} = \frac{\pi}{\sqrt{2M}} \qquad h_s = \left(\frac{\pi d}{\alpha_s M}\right)^{1/2} = \frac{\pi}{\sqrt{M}}$$

M	$\|E_S(h)\|$	$\|E_U(h_u)\|$	$\|E_S^s(h)\|$	$\|E_S^s(h_s)\|$	$\|E_U^s(h_u)\|$
4	.149 − 1	.168 − 1	.148 − 1	.192 − 2	.386 − 2
8	.451 − 2	.154 − 2	.454 − 2	.316 − 3	.364 − 3
16	.608 − 3	.833 − 4	.609 − 3	.161 − 4	.106 − 4
32	.262 − 4	.244 − 5	.263 − 4	.164 − 6	.622 − 7

Table 4.8 Comparison of the errors for the problem (4.55) using the nonsymmetric and the symmetric discrete systems.

The choice $h_s = (\pi d/(\alpha_s M))^{1/2}$ yields a somewhat unexpected result. A survey of Table 4.8 shows not only that the error is within the predicted asymptotic rate, but that it is somewhat better. Whereas the asymptotic equality indicates that the accuracy for the selection $h_s = (\pi d/(\alpha_s M))^{1/2}$ should be about the same as that for the choice $h = (\pi d/(\alpha M))^{1/2}$, there is no reason to expect any increase in accuracy. Based on these numerical computations, one might conjecture that the choice $h_s = (\pi d/(\alpha_s M))^{1/2}$ always leads to this increased accuracy. For problems on $(0, 1)$, this may well be the case. However, numerical results indicate that this unexpected accuracy may be map-dependent [4]. Finally, it should be pointed out that this same problem was computed with the system in Example 4.8 and $h_s = \pi/\sqrt{M}$. These results show no increase in accuracy.

∎

Example 4.13 Consider approximating the solution of

$$-u''(x) + \frac{3}{4x^2}u(x) = \sqrt{x}e^{-x}(3-x), \quad 0 < x < \infty \qquad (4.72)$$

$$u(0) = \lim_{x \to \infty} u(x) = 0$$

which is given by $u(x) = x^{3/2}e^{-x}$. The discretization is given by

$$\left\{\frac{-1}{h^2}I^{(2)} + I^{(0)}\right\} D\left(\frac{1}{\sqrt{x}}\right) \vec{u} = D(x^{3/2})\vec{f}.$$

Selecting $\alpha = 3/2$ gives the mesh size $h = \pi/\sqrt{3M}$. The definition for N in (4.28) with $\beta = 1$ dictates $M = 3N/2$. This can be significantly reduced for problems on $(0, \infty)$ when the map $\phi(x) = ln(x)$ is used. Mimicking the discussion following Example 3.10, this economization is achieved when the solution of (4.1) satisfies for $\beta > 0$,

$$u(x) = \mathcal{O}(e^{-\beta x}), \quad x \to \infty. \tag{4.73}$$

From (3.36) the selection for N defined by

$$N_e = \left[\left|\frac{1}{h} ln\left(\frac{\alpha}{\beta} Mh\right) + 1\right|\right] \tag{4.74}$$

is large enough to balance, and maintain, the exponential convergence rate.

Since (4.72) is a self-adjoint differential equation, the symmetric discrete system can also be used on this problem. Using (4.71) one finds $\alpha_s = \alpha - 1/2 = 1$ so that from (4.68) with $d = \pi/2$

$$h_s = \pi/\sqrt{2M}. \tag{4.75}$$

As remarked above, one can reduce the size of the system via the analogue of (4.74). If N is replaced by (4.74) with $\alpha_s = 1$, $\beta = 1$, and h_s from (4.75), the choice for N becomes

$$N_e = \left[\left|\frac{\sqrt{2M}}{\pi} ln\left(\pi\sqrt{\frac{M}{2}}\right) + 1\right|\right]. \tag{4.76}$$

The results in Table 4.9 are roughly the same for N_e in (4.76) (reported as $\|E_S^s(h_s)\|$) or $N = M$. This is in spite of the difference in the size of the system (4.61) corresponding to the two different choices N_e or N. The matrix sizes are 7, 12, 21, and 40 for the former selection and 9, 17, 33, and 65 for the latter. This economization plays a far more significant role in the Sinc-Galerkin solution of time-dependent parabolic problems than it does in the present example. However, the economization of a method is always welcome, especially when it requires little or no effort on the user's part.

M	N_e	$h_s = \pi/\sqrt{2M}$	$\|E_S^s(h_s)\|$
4	2	1.111	$.489 \times 10^{-1}$
8	3	.785	$.812 \times 10^{-2}$
16	4	.555	$.437 \times 10^{-3}$
32	7	.393	$.793 \times 10^{-5}$

Table 4.9 Economization of the Sinc-Galerkin system for the problem (4.72) using the $ln(x)$ map.

■

There is a discretization corresponding to first-order problems that is conveniently derived from the system (4.39). To obtain this approximation, consider the initial value problem

$$u'(x) = f(x), \quad 0 < x < \infty \qquad (4.77)$$
$$u(0) = 0$$

where the behavior of the solution $u(x)$ at infinity will be addressed below. If in the system (4.39) the matrices arising from the u'' term are deleted, $p(x) \equiv 1$, and $q(x) \equiv 0$, then the system takes the form

$$\left[\frac{-1}{h} I^{(1)} - D\left(\frac{w'}{\phi'w}\right)\right] D(w)\vec{u} = D\left(\frac{w}{\phi'}\right)\vec{f} \qquad (4.78)$$

for an arbitrary weight function $w(x)$. Just as the matrix $\mathcal{A}(w)$ in (4.40) was isolated as a distinguished portion of the system (4.39), the matrix

$$\mathcal{B}(w) = \frac{-1}{h} I^{(1)} - D\left(\frac{w'}{\phi'w}\right) \qquad (4.79)$$

has an equally prominent role in the system (4.78). The system (4.78) is the one obtained by orthogonalizing the residual associated with (4.77), where the assumed approximate solution of (4.77) is defined by

$$u_m(x) = \sum_{k=-M}^{N} u_k S(k, h) \circ (\phi(x)). \qquad (4.80)$$

THE SINC-GALERKIN METHOD   137

The development leading to the system (4.78), as well as to (4.30), uses the fact that the solution $u(x)$ vanishes at both endpoints. Hence, if there is to be any hope that $u_m(x)$ in (4.80) will provide an accurate approximation to the true solution of (4.77), one must deal with the implicit assumption that $u(\infty) = 0$. There are, however, situations when this assumption is free of charge. The heat equation subject to Dirichlet boundary conditions is one such instance. Here, the case when the implicit assumption is in force will be discussed first followed by the case when the behavior of the solution at infinity is not known.

For reasons that will emerge in Section 6.1 (the eigenvalues of $\mathcal{B}(w)$ have positive real part), select the weight

$$w(x) = (\phi'(x))^{1/2} \tag{4.81}$$

so that when $\phi(x) = \ell n(x)$,

$$\mathcal{B}(\sqrt{\phi'}) = \frac{-1}{h} I^{(1)} + \frac{1}{2} I^{(0)}. \tag{4.82}$$

Note that if (4.77) had been the problem

$$u'(x) + q(x)u(x) = f(x), \quad 0 < x < \infty$$
$$u(0) = 0 \tag{4.83}$$

then the only change in the discrete system (4.78) would have been the addition of a diagonal matrix (see (4.30)) so that the system for (4.83) takes the form

$$\left[\mathcal{B}(w) + D\left(\frac{q}{\phi'}\right)\right] D(w)\vec{u} = D\left(\frac{w}{\phi'}\right) \vec{f}. \tag{4.84}$$

The following example illustrates this method for a first-order initial value problem whose solution vanishes at infinity.

Example 4.14 The discretization for the problem on $(0, \infty)$ given by

$$u'(x) + \frac{3}{2x} u(x) = \sqrt{x}(3-x)e^{-x}, \quad 0 < x < \infty \tag{4.85}$$
$$u(0) = 0$$

(which has solution $u(x) = x^{3/2}e^{-x}$) takes the form (from (4.84) with the weight in (4.81))

$$\left[\frac{-1}{h}I^{(1)} + 2I^{(0)}\right]D\left(\frac{1}{\sqrt{x}}\right)\vec{u} = D\left(\sqrt{x}\right)\vec{f}.$$

The solution $\vec{u} = (u_{-M}, \ldots, u_0, \ldots, u_N)^T$ gives the coefficients in the approximate Sinc-Galerkin solution

$$u_m(x) = \sum_{k=-M}^{N} u_k S(k,h) \circ (\ell n(x)).$$

The parameters α, β, h, and N_e for (4.85) are the same as those used in Example 4.13. This follows from (4.30) and (4.31). In Table 4.10, $\alpha = 3/2$, $\beta = 1$, $h = \pi/\sqrt{3M}$. The value N_e is given by (4.74).

M	N_e	$h = \pi/\sqrt{3M}$	$\|E_S(h)\|$
4	1	0.907	$.415 \times 10^{-2}$
8	2	0.641	$.394 \times 10^{-2}$
16	3	0.453	$.566 \times 10^{-3}$
32	6	0.321	$.322 \times 10^{-4}$

Table 4.10 Sinc-Galerkin errors for the first-order initial value problem (4.85).

■

In the event that the steady-state solution of (4.83) is nonzero, one may alter the form of the approximate solution to read

$$u_A(x) = \sum_{k=-M}^{N} u_k S(k,h) \circ (\ell n(x)) + u_{N+1}\omega_\infty(x) \quad (4.86)$$

where the function $\omega_\infty(x)$ is defined by

$$\omega_\infty(x) = \frac{x}{x+1} \quad (4.87)$$

and $u_{N+1} \approx u(\infty)$. The only alteration to the discrete system (4.84) is a bordering of the coefficient matrix. The approximate solution (4.86) is orthogonalized with respect to an extra sinc basis function $S(N+1,h) \circ (\ell n(x))$ to generate the discrete system for (4.83), which is given by

$$\mathcal{B}\vec{u} = D\left(\frac{w}{\phi'}\right)\vec{f}. \qquad (4.88)$$

The $(m+1) \times (m+1)$ matrix $\mathcal{B}$ is given by

$$\mathcal{B} \equiv \left[\tilde{\mathcal{B}} \mid \vec{b}_{N+1}\right] \qquad (4.89)$$

where the k-th component of the $(m+1) \times 1$ vector $\vec{b}_{N+1}$ is

$$(\vec{b}_{N+1})_k = (\omega'_\infty + q\omega_\infty)\left(\frac{w}{\phi'}\right)(x_k).$$

One way to construct the matrix $\tilde{\mathcal{B}}$ is to multiply the $(m+1) \times m$ matrix

$$\mathcal{B}_{ns} = \frac{-1}{h} I^{(1)} - I^{(0)} D\left(\frac{w' - qw}{\phi' w}\right) \qquad (4.90)$$

by the $m \times m$ diagonal matrix $D(w)$. Specifically, obtain the $(m+1) \times m$ matrices $I^{(0)}$ and $I^{(1)}$ by letting the indices in the definition of $\delta_{jk}^{(0)}$ in (4.14) and of $\delta_{jk}^{(1)}$ in (4.15) run from $j = -M, \ldots, N+1$ and $k = -M, \ldots, N$. The $(m+1) \times m$ "identity" $I^{(0)}$ is simply an $m \times m$ identity with a row of zeros appended as the bottom row.

Example 4.15 A problem with a nonzero steady state is given by

$$u'(x) - u(x) = -1 + 2e^{-x}, \quad 0 < x < \infty$$
$$u(0) = 0. \qquad (4.91)$$

The solution of the system

$$\mathcal{B}\vec{u} = D(\sqrt{x})\vec{f}$$

with $\mathcal{B}$ given by (4.89) ($q(x) = 1$), $\mathcal{B}_{ns}$ given by (4.90), and

$$(\vec{b}_{N+1})_k = \left(\frac{1}{(x_k+1)^2} - \frac{x_k}{x_k+1}\right)x_k^{1/2}$$

gives the coefficients $\vec{u} = (u_{-M}, \ldots, u_{N+1})^T$ in the Sinc-Galerkin approximation (4.86) to the true solution of (4.91), $u(x) = 1 - e^{-x}$. Select $\alpha = \beta = 1$ so that $h = \pi/\sqrt{2M}$. In contrast to the selection N_e in Example 4.14, select $N = M$ since the solution does not go to zero as x goes to infinity. The last column of Table 4.11 includes in the error measurement $\|E_S(h)\|$ the quantity $\|u_{N+1} - u(\infty)\|$. Thus the coefficient u_{N+1} is an accurate approximation of $u(\infty) = 1$.

$M = N$	$h = \pi/\sqrt{2M}$	$\|E_S(h)\|$
4	1.111	$.202 \times 10^{-1}$
8	0.785	$.135 \times 10^{-2}$
16	0.555	$.215 \times 10^{-4}$
32	0.393	$.150 \times 10^{-5}$

Table 4.11 Sinc-Galerkin errors for the approximate steady-state solution of (4.91).

∎

One could apply the procedure of Example 4.15 to the initial value problem in Example 4.14. That is, in the event that one does not know a priori whether the solution of the initial value problem is zero at the terminal point (infinity in the above case), one introduces the boundary function $\omega_\infty(x)$ into the approximate solution and treats the problem as a boundary value problem. This procedure is carried out for a parabolic partial differential equation in Example 6.5. A sort of shooting method that just takes one shot!

4.4 Treatment of Boundary Conditions

In the previous section the development of the Sinc-Galerkin technique for homogeneous Dirichlet boundary conditions provided the most convenient approach since, by construction, the sinc functions composed with the various conformal maps, $S(j,h) \circ \phi$, are zero at the endpoints of the interval. If the boundary conditions are nonhomogeneous Dirichlet, then a straightforward transformation converts these conditions to the homogeneous Dirichlet conditions and the results of Section 4.3 apply (Exercise 4.7). If the boundary data is given by radiation conditions, whether homogeneous or nonhomogeneous, a modification of the previous approach is needed. In the nonhomogeneous case a change of variables is again used to transform the problem to one that involves only homogeneous boundary conditions (Exercise 4.6). Hence the focus in this section will be on the homogeneous case.

The development implicitly assumes that the problem is stated for a finite interval, although, as illustrated by Example 4.15 of the last section, the modification requisite for different intervals simply involves a change in the form of the assumed boundary functions. Indeed, the matrix developed in this section (given in (4.96)) with the appropriate map ϕ is the system one would use in conjunction with an adjustment of boundary basis functions for other intervals.

Equipped with the above comments, it is sufficient to address the following problem

$$\begin{aligned} Lu(x) \equiv -u''(x) &\quad+\quad p(x)u'(x) + q(x)u(x) = f(x), \quad a < x < b \\ a_0 u(a) &\quad-\quad a_1 u'(a) = 0 \\ b_0 u(b) &\quad+\quad b_1 u'(b) = 0 \end{aligned} \qquad (4.92)$$

where it is assumed that the coefficients in the boundary conditions are nonnegative and not both a_1 and b_1 are zero (the latter is spelled out in Exercise 4.7).

The treatment for any problem involving derivatives of u in the boundary conditions is simply a weighted version of the development in Section 4.2. Since $\frac{d}{dx}[S(k,h) \circ \phi(x)]$ is undefined at $x = a$ and b (recall Example (3.18)), and because the mixed boundary conditions must be handled at the endpoints by the approximate solution, a weight is introduced into this approximate solution as follows:

$$u_A(x) = c_{-M-1}\omega_a(x) + \sum_{k=-M}^{N} c_k \xi_k(x) + c_{N+1}\omega_b(x). \qquad (4.93)$$

In (4.93)

$$\xi_k(x) = S(k,h) \circ \phi(x)(g/\phi'(x)), \quad k = -M, \ldots, N \qquad (4.94)$$

where g plays the role of a weight adjustment that may be altered according to the various boundary conditions in (4.92). The boundary basis functions ω_a and ω_b are cubic Hermite functions given by

$$\omega_a(x) = a_0 \frac{(x-a)(b-x)^2}{(b-a)^2} + a_1 \frac{(2x+b-3a)(b-x)^2}{(b-a)^3}$$

and

$$\omega_b(x) = b_1 \frac{(-2x+3b-a)(x-a)^2}{(b-a)^3} + b_0 \frac{(b-x)(x-a)^2}{(b-a)^2}.$$

These boundary functions interpolate the boundary conditions in (4.92) via the identities

$$\omega_a(a) = a_1, \ \omega_a'(a) = a_0$$
$$\omega_b(b) = b_1, \ \omega_b'(b) = -b_0$$

and

$$\omega_a(b) = \omega_a'(b) = 0$$
$$\omega_b(a) = \omega_b'(a) = 0.$$

The middle term on the right-hand side of (4.93)

$$u_m(x) = \sum_{k=-M}^{N} c_k \xi_k(x), \quad m = M + N + 1$$

is a weighted version of the assumed approximate solution used in Section 4.2.

There are $m + 2$ unknown coefficients in (4.93) and these are determined by orthogonalizing the residual

$$\left(Lu - f, \frac{S(j,h) \circ \phi \, g}{\phi'}\right) = 0, \quad j = -M-1, \ldots, N+1$$

with respect to the $m + 2$ basis functions $(gS(j,h) \circ \phi)/\phi'$, where the inner product is the weighted inner product introduced at the

THE SINC-GALERKIN METHOD

beginning of Section 4.2. It is more in the spirit of Galerkin to orthogonalize with respect to the expansion base, but due to the accurate point evaluation quadratures, the Petrov–Galerkin approach used here is both accurate and more convenient.

Writing out these inner products gives

$$-\int_a^b u''(x)\frac{S(j,h)\circ\phi(x)}{\phi'(x)}g(x)w(x)dx$$
$$+\int_a^b p(x)u'(x)\frac{S(j,h)\circ\phi(x)}{\phi'(x)}g(x)w(x)dx$$
$$+\int_a^b q(x)u(x)\frac{S(j,h)\circ\phi(x)}{\phi'(x)}g(x)w(x)dx$$
$$=\int_a^b f(x)\frac{S(j,h)\circ\phi(x)}{\phi'(x)}g(x)w(x)dx.$$

It was mentioned in Section 4.2 that the same discrete system is obtained whether the integration by parts is performed on the above or whether the assumed approximate solution u_A is substituted into the above and the integration by parts is performed subsequently. In the present case, the latter is the method of choice due to the boundary basis. Specifically, replace u by u_A, where the terms in u_A involving each of c_{-M-1} and c_{N+1} are moved to the right-hand side of the equation. Setting

$$\tilde{f}(x) = f(x) - c_{-M-1}\omega_a(x) - c_{N+1}\omega_b(x)$$

and

$$\tilde{w}(x) = \frac{g(x)w(x)}{\phi'(x)}$$

the equation now reads

$$-\int_a^b u_m''(x)S(j,h)\circ\phi(x)\tilde{w}(x)dx$$
$$+\int_a^b p(x)u_m'(x)S(j,h)\circ\phi(x)\tilde{w}(x)dx$$
$$+\int_a^b q(x)u_m(x)S(j,h)\circ\phi(x)\tilde{w}(x)dx$$
$$=\int_a^b \tilde{f}(x)S(j,h)\circ\phi(x)\tilde{w}(x)dx.$$

These inner products are identical to the inner products in Theorem 4.4 if $u_m = u$ and $\tilde{w} = w$. Using those results gives

$$\sum_{k=-M}^{N} \left[\frac{-1}{h^2} \delta_{jk}^{(2)} \phi'(x_k)\tilde{w}(x_k) \right.$$
$$\left. - \frac{1}{h} \delta_{jk}^{(1)} \left(\frac{\phi''(x_k)\tilde{w}(x_k)}{\phi'(x_k)} + 2\tilde{w}'(x_k) \right) - \delta_{jk}^{(0)} \frac{\tilde{w}''(x_k)}{\phi'(x_k)} \right] u_m(x_k)$$
$$- \sum_{k=-M}^{N} \left[\frac{1}{h} \delta_{jk}^{(1)} p(x_k)\tilde{w}(x_k) + \delta_{jk}^{(0)} \frac{(p\tilde{w})'(x_k)}{\phi'(x_k)} \right] u_m(x_k)$$
$$+ \sum_{k=-M}^{N} \delta_{jk}^{(0)} \frac{q(x_k)\tilde{w}(x_k)}{\phi'(x_k)} u_m(x_k)$$
$$= \sum_{k=-M}^{N} \delta_{jk}^{(0)} \frac{\tilde{f}(x_k)\tilde{w}(x_k)}{\phi'(x_k)}, \quad j = -M-1, \ldots, N+1.$$

Restoring the nonhomogeneous term f, using (4.93) and (4.94) to write $u_m(x_k) = c_k \xi(x_k) = c_k g(x_k)/\phi'(x_k)$ and moving all unknowns to the left-hand side of the equation gives

$$\frac{L(\omega_a)(x_j)\tilde{w}(x_j)}{\phi'(x_j)} c_{-M-1} + \sum_{k=-M}^{N} \left[\frac{-1}{h^2} \delta_{jk}^{(2)} \phi'(x_k)\tilde{w}(x_k) \right.$$
$$- \frac{1}{h} \delta_{jk}^{(1)} \left(\frac{\phi''(x_k)}{\phi'(x_k)} \tilde{w}(x_k) + 2\tilde{w}'(x_k) \right.$$
$$\left. \left. p(x_k)\tilde{w}(x_k) \right) \right] \left(\frac{g(x_k)}{\phi'(x_k)} \right) c_k \qquad (4.95)$$
$$- \left(\frac{\tilde{w}''(x_j) + (p\tilde{w})'(x_j) - q(x_j)\tilde{w}(x_j)}{\phi'(x_j)} \right) \left(\frac{g(x_j)}{\phi'(x_j)} \right) c_j$$
$$+ \frac{L(\omega_b)(x_j)\tilde{w}(x_j)}{\phi'(x_j)} c_{N+1} = \frac{f(x_j)\tilde{w}(x_j)}{\phi'(x_j)}.$$

The maneuvering in the preceding steps was done for a couple of reasons. By defining $\tilde{f}$, the inner products were not only set up so that the previous work applied but also an integration by parts is avoided in the boundary functions. Letting $j = -M-1, \ldots, N+1$ gives the $(m+2) \times m$ matrix $\tilde{A}$ defined by

$$\tilde{A} = A_{ns} D\left(\frac{g}{\phi'} \right)$$

where the $(m+2) \times m$ nonsquare matrix A_{ns} is given by

$$A_{ns} \equiv \left\{ \frac{-1}{h^2} I^{(2)} D(\phi' \tilde{w}) - \frac{1}{h} I^{(1)} D\left(\frac{\phi''}{\phi'} \tilde{w} + 2\tilde{w}' + p\tilde{w} \right) \right.$$
$$\left. - I^{(0)} D\left(\frac{\tilde{w}'' + (p\tilde{w})' - q\tilde{w}}{\phi'} \right) \right\}. \quad (4.96)$$

This matrix is the $(m+2) \times m$ analogue of the matrix $\mathcal{A}(1/\phi')$ in (4.40). The matrices $I^{(\ell)}$, $\ell = 0, 1, 2$ are $(m+2) \times m$ and the diagonal matrices are $m \times m$. The discussion preceding Example 4.15 applies verbatim to the construction of $\tilde{A}$.

With both g and w unspecified, there are a number of options available for the final discrete system, but matters are not quite as open as they may at first appear. The function g must leave the assumed approximate solution u_A differentiable at one or both endpoints. The weight function w used in the inner product has been implicitly assumed to have the property that in the integration by parts the boundary integrals vanish. A restriction that proves useful due both to convenience and established properties of matrix performance in the homogeneous Dirichlet problem sets

$$g(x)w(x) = 1.$$

In this case $\tilde{w} = 1/\phi'$, and the matrix A_{ns} is the $(m+2) \times m$ analogue of the coefficient matrix in (4.43). Now select g and w to balance the competing goals of differentiability of u_A versus vanishing of the boundary terms in the integration by parts. A selection that will handle problems (4.92) with both a_1 and b_1 nonzero takes $g = w = 1$. Then the base matrix $\tilde{A}$ is a diagonal multiple of (4.43). This matrix is used in Example 4.16.

In general, the entire discrete system in (4.95) is constructed as follows. Define the components of the $(m+2) \times 1$ column vectors by

$$(\vec{a}_{-M-1})_j = \frac{L(\omega_a)(x_j)\tilde{w}(x_j)}{\phi'(x_j)},$$

$$(\vec{a}_{N+1})_j = \frac{L(\omega_b)(x_j)\tilde{w}(x_j)}{\phi'(x_j)}$$

and
$$\vec{b}_j = \frac{f(x_j)\tilde{w}(x_j)}{\phi'(x_j)}.$$

Then the discrete system (4.95) for (4.92) is represented as

$$\mathcal{A}\vec{c} = \vec{b} \qquad (4.97)$$

where the $(m+2) \times (m+2)$ matrix A is the bordered matrix

$$\mathcal{A} = \begin{bmatrix} \vec{a}_{-M-1} \mid \tilde{A} \mid \vec{a}_{N+1} \end{bmatrix} \qquad (4.98)$$

and $\vec{c} = (c_{-M-1}, \ldots, c_{N+1})^T$ are the coefficients in the approximate solution (4.93).

In the special case mentioned above ($g(x) \equiv w(x) \equiv 1$), the matrix A_{ns} in (4.96) takes the familiar form

$$A_{ns} = \left\{ \frac{-1}{h^2} I^{(2)} + \frac{1}{h} I^{(1)} D\left(\frac{\phi''}{(\phi')^2} - \frac{p}{\phi'}\right) \right. \\ \left. + D\left(\frac{-1}{\phi'}\left(\frac{1}{\phi'}\right)'' - \frac{1}{\phi'}\left(\frac{p}{\phi'}\right)' + \frac{q}{(\phi')^2}\right) \right\} \qquad (4.99)$$

while

$$(\vec{a}_{-M-1})_j = \frac{L(\omega_a)(x_j)}{(\phi'(x_j))^2},$$

$$(\vec{a}_{N+1})_j = \frac{L(\omega_b)(x_j)}{(\phi'(x_j))^2}$$

and

$$\vec{b}_j = \frac{f(x_j)}{(\phi'(x_j))^2}.$$

The solution $\vec{c}$ to (4.97) gives the coefficients in the approximate Sinc-Galerkin solution

$$u_A(x) = c_{-M-1}\omega_a(x) + \sum_{k=-M}^{N} c_k \frac{S(k,h) \circ \phi(x)}{\phi'(x)} + c_{N+1}\omega_b(x).$$

THE SINC-GALERKIN METHOD

Example 4.16 The differential equation

$$-v''(x) + \frac{1}{x} v'(x) + \frac{1}{x^2} v(x) = f(x)$$
$$v(0) - 2v'(0) = 1$$
$$3v(1) + v'(1) = 9$$

where $f(x) = \frac{\sqrt{x}}{4}(-41x^2 + 34x - 1) - 2x + \frac{1}{x^2}$, has the solution given by $v(x) = x^{5/2}(1-x)^2 + x^3 + 1$. This may be transformed into the following form with homogeneous boundary conditions

$$-u''(x) + \frac{1}{x} u'(x) + \frac{1}{x^2} u(x) = \bar{f}(x)$$
$$u(0) - 2u'(0) = 0 \quad (4.100)$$
$$3u(1) + u'(1) = 0$$

where

$$\bar{f}(x) = f(x) - \left(\frac{3}{5x} + \frac{3x+11}{5x^2}\right)$$

as is shown in Exercise 4.6. The $(m+2) \times m$ matrix A_{ns} in (4.99) is

$$A_{ns} = \frac{-1}{h^2} I^{(2)} + \frac{1}{h} I^{(1)} D\left((2x-1)\right) + I^{(0)} D\left((1-x)(2x+1)\right).$$

To determine the quantities α and β, one notes that $u_m(x)$ is approximating the homogeneous portion of u, which is given by $x^{5/2}(1-x)^2$. Since the error in truncating the sums in the inner products in Section 4.2 bounded this homogeneous solution with α and β given by (4.51), it follows that $\alpha = 5/2$ and $\beta = 2$. The only alteration for the development in this section is the division by $\phi'(x)$ in the basis for the approximate solution $u_A(x)$. The net effect of this division is to subtract one from each of α and β ($1/\phi'(x) = x(1-x)$ multiplies the approximate solution). All other selections are the same as previously defined. Hence with $\alpha = 3/2$ and $\beta = 1$, put $N = 3M/2$ and $h = \pi/\sqrt{3M}$. These are the selections made to generate Table 4.12.

M	N	$h = \pi/(\sqrt{3M})$	$\|E_S(h)\|$	$\|E_S(\tilde{h})\|$
4	6	0.907	$.248 \times 10^{-1}$	$.356 \times 10^{-1}$
8	12	0.641	$.731 \times 10^{-2}$	$.167 \times 10^{-1}$
16	24	0.453	$.806 \times 10^{-3}$	$.338 \times 10^{-2}$
32	48	0.321	$.243 \times 10^{-4}$	$.228 \times 10^{-3}$

Table 4.12 Errors in the Sinc-Galerkin solution of (4.100) with the "optimal" mesh h versus the layman's mesh $\tilde{h}$.

It is a little optimistic to hope for a nonpedagogical problem that one would know these "optimal" parameters explicitly. What one does in practice is set $\alpha = \beta = 1$ and select h according to (4.29) and $N = M$. The reason behind this is that the solution of the boundary value problem and its derivative are defined at the boundary ($a = 0$ and $b = 1$). Hence, its homogeneous part vanishes to some order. This minimal order is at least one so that provides a layman's choice. In Table 4.12 these parameters with $\tilde{h} = \pi/\sqrt{2M}$ are used on this problem. The results are listed in the final column under the heading $\|E_S(\tilde{h})\|$. Note that there is roughly a digit's difference in the error corresponding to the two choices h and $\tilde{h}$.

■

As a final note here on selection availability, if in (4.92) $a_1 = 0$, then there is no need for the introduction of the base function $w_a(x)$. Correspondingly, there is little point in weighting the approximate solution at the left endpoint a. In this case, select $g(x) = 1/x$ and $w(x) = x$. With $g(x)w(x) = 1$ in force, the resulting system takes the form of (4.98) with the first column removed and the diagonal multiplier replaced by $D(1-x)$. This selection is used in the following example.

Example 4.17 The advection problem

$$\begin{aligned} -u''(x) + \kappa u'(x) &= f(x) \\ u(0) &= 0 \\ \rho u(1) + u'(1) &= 0 \end{aligned} \qquad (4.101)$$

with $\rho = \infty$ (that is $u(1) = \lim_{\rho \to \infty}(-u'(1)/\rho) = 0$) was computed in Example 4.6. The true solution to the stated problem is given by

$$u_\rho(x) = \frac{\rho+1}{(\kappa+\rho)e^\kappa - \rho}(1 - e^{\kappa x}) + x$$

and the approximate solution takes the form

$$u_A(x) = \sum_{j=-M}^{M} c_j S(j,h) \circ \phi(x)(1-x) + c_{M+1}\omega_1(x)$$

with the changes made as suggested above ($g(x) = 1/x$ and $w(x) = x$ in (4.97)). Figure 4.2 is a graph of the numerical solution $u_A(x)$ corresponding to $M = N = 16$; $\kappa = 10$; and $\rho = 0, 10, 100, \infty$.

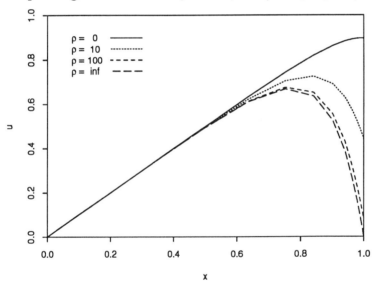

Figure 4.2 Computed solution of (4.101) with $M = N = 16$; $\kappa = 10$; and $\rho = 0, 10, 100, \infty$.

The boundary basis function ω_1 is

$$\omega_1(x) = x^2(1 - \gamma x)$$

where

$$\gamma = \frac{2\rho+1}{3\rho+1}.$$

The system to solve for this problem is the same as that in (4.96) with the diagonal multiplier there replaced by $D(1-x)$ and only one border on the matrix $\mathcal{A}$. For $\rho = 10$, Table 4.13 gives the numerical results for $\kappa = 10$ and $\kappa = 100$.

Note that one could also take $\omega_1(x)$

$$\omega_1(x) = x - \gamma x^2$$

where

$$\gamma = \frac{\rho+1}{\rho+2}$$

since the derivative of ω_1 need not be zero at $x = 0$. This is simpler but is applicable only to the specific boundary conditions in (4.101).

$M = N$	$h = \pi/(\sqrt{2M})$	$\|E_S(h)\|$
\multicolumn{3}{c}{$\kappa = 10$}
4	1.111	$.230 \times 10^{+0}$
8	0.785	$.885 \times 10^{-1}$
16	0.555	$.916 \times 10^{-2}$
32	0.393	$.512 \times 10^{-3}$
\multicolumn{3}{c}{$\kappa = 100$}
4	1.111	$.424 \times 10^{+0}$
8	0.785	$.164 \times 10^{+0}$
16	0.555	$.272 \times 10^{-1}$
32	0.393	$.122 \times 10^{-2}$

Table 4.13 Errors in the Sinc-Galerkin solution of (4.101) for $\kappa = 10$ and $\kappa = 100$ with $\rho = 10$.

4.5 Analytic Considerations

Several items in the earlier sections have not been verified. For example, the invertibility of (4.43) or (4.61) was only assumed. In Example 4.5, it was remarked that the parameters α and β could be chosen via a knowledge of the true solution. Whereas one does not need optimal parameters for a good approximation (Example 4.16), if one can conveniently obtain such parameters, then there is less computational work involved. These items, along with guidelines for the map selection will be given in this section.

The various properties of $I^{(1)}$ and $I^{(2)}$ defined in (4.15) and (4.16), and displayed in (4.37) and (4.38), respectively, are given first.

Theorem 4.18 *The $m \times m$ matrix $I^{(2)}$ is a symmetric, negative definite Toeplitz matrix (hence centrosymmetric). If its eigenvalues are denoted by $\{-\lambda_j^{(2)}\}_{j=-M}^{N}$, then*

$$0 < 4\sin^2\left(\frac{\pi}{2(m+1)}\right) \leq \lambda_{-M}^{(2)} \leq \ldots \leq \lambda_N^{(2)} \leq \pi^2. \quad (4.102)$$

The spectral condition number $\kappa(I^{(2)}) \equiv \|I^{(2)}\|_2 \, \|(I^{(2)})^{-1}\|_2$ of $I^{(2)}$ is bounded by

$$\kappa(I^{(2)}) \leq \frac{\pi^2}{4\sin^2\left(\frac{\pi}{2(m+1)}\right)} = \mathcal{O}((m+1)^2), \quad m \to \infty.$$

Proof Exercise 4.4 shows that the matrix $I^{(2)}$ is the finite Toeplitz form obtained from $f(t) = -t^2$ on $[-\pi, \pi]$. The lower bound in (4.102) is obtained (see Theorem A.19) by comparing $I^{(2)}$ with the tridiagonal Toeplitz matrix $T = tri(-1, 2, -1)$ (Definition A.15) of the same order ($m \times m$). The latter matrix is the Toeplitz form derived from $g(t) = 2 - 2\cos(t)$. The eigenvalues of T are $\mu_j = 4\sin^2\left(\frac{j\pi}{2(m+1)}\right)$, $j = 1, 2, \ldots, m$. Now note that $-f(t) \geq g(t)$. Due to symmetry, the spectral condition number $\kappa(I^{(2)}) = \lambda_N^{(2)}/\lambda_{-M}^{(2)} \leq \pi^2 / \left(4\sin^2\left(\frac{\pi}{2(m+1)}\right)\right)$ from (4.102). ∎

Theorem 4.19 *The $m \times m$ matrix $I^{(1)}$ is a skew-symmetric Toeplitz matrix. If its eigenvalues are denoted by $\{i\lambda_j^{(1)}\}_{j=-M}^{N}$ then*

$$-\pi \leq \lambda_{-M}^{(1)} \leq \cdots \leq \lambda_N^{(1)} \leq \pi. \qquad (4.103)$$

Proof The matrix $I^{(1)}$ is the finite Toeplitz form obtained from $f(t) = -it$ on $[-\pi, \pi]$. See Exercise 4.5.

∎

The invertibility of the system in (4.43) depends both on the nature of the coefficients q and p in the differential equation (4.1) and on a continuity argument (as $h \to 0$). Whether this approach to zero is compatible with the selection of h in (4.29) is not a consequence of the continuity argument. The authors leave this (somewhat) unsatisfactory state of affairs with the numerical reassurance that they have never encountered any difficulty in solving the system in (4.43).

More can be said about the solution of the discrete system in (4.61). This is due to the absence of the matrix $I^{(1)}$. For all the maps listed in Table 4.7 (except $\phi(x) = \sinh^{-1}(x)$), the quantity $(1/(\phi')^{3/2}) \cdot (1/\sqrt{\phi'})''$ is nonpositive so that if q in (4.59) is nonnegative the coefficient matrix in (4.61) is positive definite and therefore invertible. The assumption that q is nonnegative is not extremely restrictive and can often be relaxed.

The selection of the parameters α and β can often be based on the Frobenius form of the solution near the left-hand endpoint a, where $|a| < \infty$ with a similar analysis applicable at b if b is finite. If $b = \infty$, the WKB method [6] or the various procedures described in [1] may prove useful if the optimal parameter β is desired. In practice, if the solution decays exponentially, assign $\beta = 1$ and select N_e as in (4.74). In the event that exponential decay is in question, simply assign $\beta = 1$ and select N based on (4.29).

The Frobenius considerations necessary to find α for the problem (4.59) when a is finite are sorted out in what follows.

Theorem 4.20 *Assume that $q(x) \geq 0$ on (a, b) in (4.59) and that*

$$\lim_{x \to a^+} (x - a)^2 q(x) = q_0 \geq 0.$$

THE SINC-GALERKIN METHOD

The indicial equation of (4.59) *is*

$$r(r-1) - q_0 = 0.$$

The roots of this equation are

$$r_1 = \frac{(1 + \sqrt{1+4q_0})}{2} \geq 1 \quad \text{and} \quad r_2 = \frac{(1 - \sqrt{1+4q_0})}{2} \leq 0.$$

If $f(x) = \mathcal{O}((x-a)^\gamma)$ *as* $x \to a^+$, *the solution of* (4.59) *satisfies* $u(x) = \mathcal{O}((x-a)^\alpha)$, *as* $x \to a^+$ *where*

$$\alpha = \min\{r_1, \gamma + 2\}. \tag{4.104}$$

Proof It is convenient to assume $a = 0$. A homogeneous basis of solutions for $L_s u = 0$ satisfies $\chi_1(x) = \mathcal{O}(x^{r_1})$ and $\chi_2(x) = \mathcal{O}(x^{r_2})$ ($x \to 0^+$) as long as $r_1 - r_2 = \sqrt{1+4q_0}$ is not a positive integer (Exercise 4.8). The Wronskian of χ_1 and χ_2 satisfies

$$W(x) = W(\chi_1, \chi_2)(x) \equiv \begin{vmatrix} \chi_1 & \chi_2 \\ \chi_1' & \chi_2' \end{vmatrix}(x) = \mathcal{O}(1), \quad x \to 0^+$$

since $r_1 + r_2 = 1$. A particular solution of $L_s u = f$ is given by the variation of parameters formula

$$u_p(x) = -\chi_1(x) \int_0^x \frac{f(t)\chi_2(t)}{W(t)} dt + \chi_2(x) \int_0^x \frac{f(t)\chi_1(t)}{W(t)} dt \tag{4.105}$$

$$= \mathcal{O}(x^{\gamma+2}) \text{ as } x \to 0^+.$$

Hence the solution of $L_s u = f$ is $u(x) = c_1 \chi_1(x) + c_2 \chi_2(x) + u_p(x)$, where $c_2 = 0$ since $u(0) = 0$ and $r_2 < 0$. Thus $u(x) = c_1 \chi_1(x) + u_p(x)$ and since $\chi_1(x) = \mathcal{O}(x^{r_1})$ as $x \to 0^+$, the choice of α follows. ∎

Corollary 4.21 *Assume that* $q \geq 0$ *on* (a,b) *in* (4.59) *and that*

$$\lim_{x \to a^+} (x-a)^2 q(x) = q_0 \geq 0.$$

If f is bounded at $x = a$, then for $0 \leq q_0 \leq 2$, the true solution of (4.59) satisfies $u(x) = \mathcal{O}((x-a)^\alpha)$ as $x \to a^+$, where the quantity α is given by
$$\alpha = r_1 = \left(1 + \sqrt{1 + 4q_0}\right)/2.$$

Proof If f is bounded, then $\gamma \geq 0$ and the result follows from $r_1 \leq 2$ if $q_0 \leq 2$.

∎

The above methodology applies to the more general (4.1). The only alteration is in the form of the indicial equation, which now includes the additional linear term $p_0 r$, where $\lim_{x \to a^+}(x-a)p(x) = p_0$.

For the various maps listed in Table 4.1 there is a choice of maps available in each of the cases that the domain of the problem is $(0, \infty)$ (or $(-\infty, \infty)$). In the latter case the indication seems to be that if the function u being approximated decreases very rapidly (exponential decrease), then mapping the problem (using $\phi(x) = \sinh^{-1}(x)$) is not necessary. That this indication proves true is the subject of [3]. Roughly speaking, if the function u to be approximated is rational, then use $\phi(x) = \sinh^{-1}(x)$; otherwise use $\phi(x) = x$ (no mapping).

In the case of the domain $(0, \infty)$ there are available the selections $\phi(x) = \ell n(x)$ and $\phi(x) = \ell n(\sinh(x))$. The discussion in Chapter 3 (Example 3.13) points to a situation (oscillatory functions) where the selection of the map $\phi(x) = \ell n(\sinh(x))$ is preferable to $\phi(x) = \ell n(x)$. In the case of the boundary value problem (4.1) the following criterion is handy. If the coefficients p, q, and f are analytic in a sector of the right half plane, then so is the solution of (4.1) and the map $\phi(x) = \ell n(x)$ is in general used. If it is not possible to prove that the solution of (4.1) is analytic in a sector, or if it is known that the solution has singularities near the positive half axis, then the map $\phi(x) = \ell n(\sinh(x))$ is used. These overall distinctions between the two maps are not foolproof, but for a wide class of problems they do lead to a successful numerical procedure. In problems where both maps are applicable, the use of either of the maps leads to similar errors. The only difference in the two procedures is in the size of the discrete system. The map $\phi(x) = \ell n(x)$ often leads to a smaller discrete system than does the map $\phi(x) = \ell n(\sinh(x))$ for equivalent accuracy (Exercise 4.10).

Exercises

Exercise 4.1 Show that B_{T_j} ($j = 1, 2$) in (4.11) and (4.13) vanish (on $(0,1)$) if $w(x) = [\phi'(x)]^{-1}$ and $\phi(x) = \ell n\left(\frac{x}{1-x}\right)$, where L in (4.1) has at worst regular singular points at $x = 0$ and $x = 1$. This means

$$\lim_{x \to 0^+} xp(x) = p_0 < \infty$$

and

$$\lim_{x \to 0^+} x^2 q(x) = q_0 < \infty$$

with similar statements at $x = 1$.

Exercise 4.2 Show that the assumptions in (4.50) can be replaced by $u \in B(D)$ as long as (4.53) and (4.54) are satisfied and similar conditions hold at b (assume $b < \infty$). Use the map $\phi(x) = \ell n\left(\frac{x-a}{b-x}\right)$.

Exercise 4.3 Take $b = +\infty$ in Theorem 4.7, $p(x) \equiv 0$, and $\phi(x) = \ell n(x)$. Assume that $u \in B(D_W)$ and that there is a positive constant γ so that

$$u(x) = \mathcal{O}(e^{-\gamma x}), \quad x \to \infty.$$

Simplify the assumption in (4.50). For example, $u\phi''/\phi'$ and $u\phi'$ belonging to $B(D_W)$ are the same assumption. Pay particular attention to the terms involving q and f.

Exercise 4.4 Let $f(t) = -t^2 \in L^2(-\pi, \pi)$ and find its Toeplitz form. This amounts to the calculation of the Fourier coefficients of the function $f(t) = -t^2$ using (A.17) and building the matrix in (A.18). Find the Toeplitz form for $f(t) = -t$. How is this related to $I^{(1)}$?

Exercise 4.5 The identities in (4.17) and (4.20) may be used to obtain $\delta_{jk}^{(p)}$ for all p and, in particular, for $p = 3$ and 4. Use (4.20) with $n = 0$ to find

$$\delta_{jk}^{(3)} \equiv h^3 \left[\frac{d^3}{d\phi^3} S(j, h) \circ \phi(x) \right]\bigg|_{x=x_k} \quad (4.106)$$

$$= \begin{cases} 0, & j = k \\ \dfrac{(-1)^{k-j}}{(k-j)^3}[6 - \pi^2(k-j)^2], & j \neq k. \end{cases}$$

Set $n = 1$ in (4.17) to find

$$\delta_{jk}^{(4)} \equiv h^4 \left[\frac{d^4}{d\phi^4} S(j,h) \circ \phi(x)\right]\bigg|_{x=x_k} \tag{4.107}$$

$$= \begin{cases} \dfrac{\pi^4}{5}, & j = k \\ \dfrac{-4(-1)^{k-j}}{(k-j)^4} [6 - \pi^2(k-j)^2], & j \neq k. \end{cases}$$

Apply the procedure in (4.12) (integrate by parts four times) to the fourth-order boundary value problem

$$u''''(x) + q(x)u(x) = f(x), \quad 0 < x < 1$$

subject to the homogeneous boundary conditions

$$u(0) = u(1) = u'(0) = u'(1) = 0$$

where the weight $w(x)$ is arbitrary (see [9], [10]). The boundary terms vanish under reasonable assumptions if one selects $w(x) = (\phi'(x))^{-l}$, where $l = 3/2$ or 2 with $\phi(x) = \ln(x/(1-x))$. In the former case, discussed in [11], show that the discrete system is

$$\left[\frac{1}{h^4} I^{(4)} - \frac{5}{2h^2} I^{(2)} + D(\gamma_q)\right] D((\phi')^{3/2}) \vec{u} = D\left(\frac{1}{(\phi')^{5/2}}\right) \vec{f} \tag{4.108}$$

where

$$\gamma_q(x) = \frac{1}{(\phi'(x))^{5/2}} \frac{d^4}{d\phi^4} \left(\frac{1}{(\phi'(x))^{3/2}}\right) + \frac{q(x)}{(\phi'(x))^4}$$

$$= \frac{9}{16} + (x(1-x))^4 q(x).$$

The vector $\vec{u}$ contains the coefficients in the approximate solution

$$u_m(x) = \sum_{k=-M}^{N} u_k S(k,h) \circ \phi(x).$$

Note that this system does not involve $I^{(3)}$. The system using $l = 2$ will involve $I^{(3)}$.

THE SINC-GALERKIN METHOD

The $m \times m$ matrix $I^{(3)}$, whose component definition is given in (4.106), has the form

$$\begin{bmatrix} 0 & -(6-\pi^2) & \cdots & \frac{(-1)^{m-1}(6-(m-1)^2\pi^2)}{(m-1)^3} \\ 6-\pi^2 & & & \vdots \\ \frac{-(6-2^2\pi^2)}{2^3} & \ddots & & 6-2^2\pi^2 \\ \vdots & & & -(6-\pi^2) \\ \frac{(-1)^m(6-(m-1)^2\pi^2)}{(m-1)^3} & \cdots & 6-\pi^2 & 0 \end{bmatrix}$$

and the $m \times m$ matrix $I^{(4)}$ has the form

$$\begin{bmatrix} \frac{\pi^4}{5} & 4(6-\pi^2) & \cdots & \frac{4(-1)^m(6-(m-1)^2\pi^2)}{(m-1)^4} \\ 4(6-\pi^2) & & & \vdots \\ \frac{-4(6-2^2\pi^2)}{2^4} & \ddots & & \frac{-4(6-2^2\pi^2)}{2^4} \\ \vdots & & & 4(6-\pi^2) \\ \frac{4(-1)^m(6-(m-1)^2\pi^2)}{(m-1)^4} & \cdots & 4(6-\pi^2) & \frac{\pi^4}{5} \end{bmatrix}$$

where the component definition is given in (4.107). Approximate the boundary value problem if $q(x) = 1/x^4$ and the solution is $u(x) = x^{5/2}(\ln(x))^3$. With $\alpha = 1$, $\beta = 3/2$, and $d = \pi/2$, select h and N as in Theorem 4.4. Calculate the approximate solution using $M = 4, 8, 16$, and 32. Make a table analogous to Table 4.2 in which $\|E_U(h_u)\|$ is replaced by the convergence rate $CR = \mathcal{O}(\exp(-\pi\sqrt{M/2}))$. Do you believe this convergence rate?

Exercise 4.6 Consider the problem
$$Lv(x) \equiv -v''(x) + p(x)v'(x) + q(x)v(x) = f(x), \quad a < x < b$$
$$a_0 v(a) - a_1 v'(a) = a_2 \qquad (4.109)$$
$$b_0 v(b) + b_1 v'(b) = b_2.$$

Define the interpolating boundary function
$$l(x) = Ax + B,$$

where
$$A = \frac{b_0 a_2 - a_0 b_2}{D},$$
$$B = \frac{(a_0 a - a_1)b_2 - (b_0 b + b_1)a_2}{D}$$

and
$$D = a_0 b_0 (a - b) - a_1 b_0 - a_0 b_1 \neq 0.$$

Now, define the linear shift
$$u(x) = v(x) - l(x)$$

and show that the problem (4.109) in the dependent variable u reads
$$Lu(x) \equiv -u''(x) + p(x)u'(x) + q(x)u(x) = \bar{f}(x), \quad a < x < b$$
$$a_0 u(a) - a_1 u'(a) = 0$$
$$b_0 u(b) + b_1 u'(b) = 0$$

where
$$\bar{f}(x) = f(x) - L(l)(x).$$

This is (4.92).

Exercise 4.7 In the case of a nonhomogeneous Dirichlet problem ($a_1 = b_1 = 0$) there are nonzero numbers γ and δ defining the boundary conditions
$$v(a) = \frac{a_2}{a_0} = \gamma$$

and
$$v(b) = \frac{b_2}{b_0} = \delta.$$

THE SINC-GALERKIN METHOD

Show that the transformation

$$u(x) = v(x) - \gamma \frac{b-x}{b-a} - \delta \frac{x-a}{b-a} \quad (4.110)$$

applied to the problem (4.109) yields the differential equation

$$Lu(x) \equiv -u''(x) + p(x)u'(x) + q(x)u(x) = \hat{f}(x)$$
$$u(a) = u(b) = 0$$

where

$$\hat{f}(x) = f(x) + \frac{\gamma - \delta}{b-a} p(x) - \frac{(\delta - \gamma)x + \gamma b - \delta a}{b-a} q(x).$$

The resulting discrete system for the coefficients $\vec{c} = (c_{-M}, \ldots, c_N)^T$ in the approximate sinc solution

$$u_A(x) = \sum_{k=-M}^{N} c_k S(k, h) \circ \phi(x) + \gamma \frac{b-x}{b-a} + \delta \frac{x-a}{b-a} \quad (4.111)$$

is exactly the system in (4.43), with f in that system replaced by $\hat{f}$. Notice that if $\delta = \gamma = 0$, then the discrete system obtained and the assumed solution (4.111) reduce to (4.39) and (4.41), respectively.

Exercise 4.8 If the quantity $r_1 - r_2 = \sqrt{1 + 4q_0}$ (the notation of Theorem 4.20) is a positive integer, then a homogeneous basis of solutions satisfies $\chi_1(x) = \mathcal{O}(x^{r_1})$, $x \to 0^+$ and

$$\chi_2(x) = k\chi_1(x)\ln(x) + x^{r_2} \sum_{j=0}^{\infty} c_j x^j$$

where k is a constant (possibly zero) and $c_0 \neq 0$. If $f(x) = \mathcal{O}(x^\gamma)$, show that (4.104) remains in force.

Exercise 4.9 Approximate the solution to the problem

$$-u''(x) + \frac{3}{4x^2}u(x) = \sqrt{x}e^{-x}(3-x), \quad 0 < x < \infty$$

$$u(0) = \lim_{x \to \infty} u(x) = 0$$

using the map $\phi(x) = \ln(\sinh(x))$. This is the problem in Example 4.13. The discrete system defined in (4.61) takes the form

$$\left\{\frac{-1}{h^2}I^{(2)} + D\left(\frac{4\cosh^2(x) - 3}{4\cosh^4(x)} + \frac{3\tanh^2(x)}{4x^2}\right)\right\}\vec{y}$$

$$= D(\tanh^{3/2}(x))\vec{f}$$

where

$$f(x) = \sqrt{x}e^{-x}(3-x)$$

and, as in (4.63),

$$\vec{y} = D\left(\frac{1}{\sqrt{\tanh(x)}}\right)\vec{u}^s.$$

Obtain the parameter selections from (4.68). One could directly fill the matrices above using the nodes

$$x_k = \ln(e^{kh} + \sqrt{e^{2kh} + 1})$$

from Table 3.7. It may help to note that a short calculation gives

$$\sinh(x_k) = e^{kh}$$

and

$$\cosh(x_k) = \sqrt{e^{2kh} + 1}.$$

Compare the system size (for equivalent accuracy) with the system size in that Example 4.13. Is there a choice N_e, analogous to (4.74), for the map $\phi(x) = \ln(\sinh(x))$?

Exercise 4.10 In the interest of equal time, solve the problem in Example 4.9 with the $\ln(x)$ map. As in Example 4.9, $\alpha = \beta = 1$ so that $h = \pi/\sqrt{2M}$ and N_e is given by (4.74). If the results seem a bit discouraging, recall Example 3.13.

References

[1] C. M. Bender and S. A. Orszag, *Advanced Mathematical Methods for Scientists and Engineers*, McGraw-Hill, Inc., New York, 1978.

[2] B. Bialecki, "Sinc-Type Approximations in H^1-Norm with Application to Boundary Value Problems," *J. Comput. Appl. Math.*, 25 (1989), pages 289–303.

[3] N. Eggert and J. Lund, "The Trapezoidal Rule for Analytic Functions of Rapid Decrease," *J. Comput. Appl. Math.*, 27 (1989), pages 389–406.

[4] J. Lund, "Symmetrization of the Sinc-Galerkin Method for Boundary Value Problems," *Math. Comp.*, 47 (1986), pages 571–588.

[5] S. G. Mikhlin, *Variational Methods in Mathematical Physics*, Pergamon Press Ltd., Oxford, 1964.

[6] F. W. J. Olver, *Asymptotics and Special Functions*, Academic Press, Inc., New York, 1974.

[7] P. M. Prenter, *Splines and Variational Methods*, John Wiley & Sons, Inc., New York, 1975.

[8] S. Schaffer and F. Stenger, "Multigrid-Sinc Methods," *Appl. Math. Comput.*, 19 (1986), pages 311–319.

[9] R. C. Smith, "Numerical Solution of Fourth-Order Time-Dependent Problems with Applications to Parameter Identification," Ph. D. Thesis, Montana State University, Bozeman, MT, 1990.

[10] R. C. Smith, G. A. Bogar, K. L. Bowers, and J. Lund, "The Sinc-Galerkin Method for Fourth-Order Differential Equations," *SIAM J. Numer. Anal.*, 28 (1991), pages 760–788.

[11] R. C. Smith, K. L. Bowers, and J. Lund, "Efficient Numerical Solution of Fourth-Order Problems in the Modeling of Flexible Structures," in *Computation and Control*, Proc. Bozeman Conf. 1988, Progress in Systems and Control Theory, Vol. 1, Birkhäuser, Boston, 1989, pages 283–297.

[12] F. Stenger, "A Sinc-Galerkin Method of Solution of Boundary Value Problems," *Math. Comp.*, 33 (1979), pages 85–109.

[13] F. Stenger, "Numerical Methods Based on Whittaker Cardinal, or Sinc Functions," *SIAM Rev.*, 23 (1981), pages 165–224.

Chapter 5

Steady Problems

There are a number of applications of the material in the preceding chapter that lend themselves rather directly to many problems arising in numerical analysis. This chapter deals with a few of these steady (time-independent) applications. An intimate connection between the Sinc-Galerkin method and a sinc-collocation scheme was advertised in Examples 2.17 and 2.22. There are a number of collocation procedures available and Section 5.1 explores their connections with companion Sinc-Galerkin schemes. A conjectured numerical equivalence of the two procedures is strongly indicated. One of these collocation schemes leads to an algorithm for the computation of the eigenvalues of Sturm–Liouville problems. This is the subject of Section 5.2. In Section 5.3 a different discretization is developed for the self-adjoint form in second-order differential equations. The final Section 5.4 develops the product formulation for the solution of the Poisson problem in two and three dimensions.

5.1 Sinc-Collocation Methods

A collocation scheme for the boundary value problem

$$Lu(x) \equiv -u''(x) + p(x)u'(x) + q(x)u(x) = f(x), \quad a < x < b \tag{5.1}$$
$$u(a) = u(b) = 0$$

is defined by the equations

$$Lu(x_j) = f(x_j), \quad a < x_j < b, \quad j = 1, 2, \ldots, m \tag{5.2}$$

where the requirements of the scheme are twofold. The first requirement is to select the m points $\{x_j\}$ in the evaluation in (5.2) and the second is to specify the approximations of the derivatives $u^{(n)}(x_j)$, $n = 1, 2$ on the left-hand side of (5.2). If the inner product in (4.3) is thought of as a duality pairing with the second argument $\chi_j(x)$ replaced by the functional $\delta_j(x) \equiv \delta(x - x_j)$ and u_N replaced by u, then, with a little license, (5.2) can be shown to be the orthogonalization of the residual in (4.3).

It follows from Theorem 3.4 that the approximation

$$u(x) \simeq u_m(x) = \sum_{k=-M}^{N} u_k S(k, h) \circ \phi(x)$$

provides an accurate approximation of the true solution u of (5.1) but gives no information about whether the derivative of u is accurately approximated by the derivative of u_m (Example 3.18). It was precisely this line of thinking that led to the integration by parts in the inner product approximations in Section 4.2. In order to find accurate sinc approximations to $u^{(n)}(x_j)$ ($n = 1, 2$) on the left-hand side of (5.2), a formal application of Theorem 3.16 is perhaps the most direct. This convenience comes with the expense of extremely stringent assumptions on the solution u of (5.1).

The procedure outlined in [24] avoids the use of Theorem 3.17. Using (4.30), (4.31), and (4.32), write the approximations

$$\int_a^b q(x)u(x)S(j, h) \circ \phi(x)w(x)dx \simeq h \frac{q(x_j)w(x_j)}{\phi'(x_j)} u_j, \tag{5.3}$$

$$\int_a^b p(x)u'(x)S(j,h)\circ\phi(x)w(x)dx \tag{5.4}$$

$$\simeq -h\sum_{k=-M}^{N}\left\{\frac{\delta_{jk}^{(1)}}{h}(pw)(x_k) + \delta_{kj}^{(0)}\left(\left(\frac{(pw)'}{\phi'}\right)(x_k)\right)\right\}u_k$$

and

$$-\int_a^b u''(x)S(j,h)\circ\phi(x)w(x)dx \simeq -h\sum_{k=-M}^{N}\left\{\frac{\delta_{jk}^{(2)}}{h^2}(\phi'w)(x_k)\right.\tag{5.5}$$

$$\left.+\frac{\delta_{jk}^{(1)}}{h}\left(\frac{\phi''}{\phi'}w + 2w'\right)(x_k) - h\left(\frac{w''}{\phi'}\right)(x_k)\right\}u_k.$$

Now, use the approximate point evaluation quadrature in (4.30) to write

$$\int_a^b F(x)w(x)S(j,h)\circ\phi(x)dx \simeq \frac{hw(x_j)}{\phi'(x_j)}F(x_j). \tag{5.6}$$

Equating the right-hand sides of (5.3), (5.4), and (5.5) with the right-hand side of (5.6), where in the latter $F = qu$, pu', and $-u''$, respectively, leads to the approximations

$$\frac{q(x_j)w(x_j)u(x_j)}{\phi'(x_j)} \simeq \frac{q(x_j)w(x_j)u_j}{\phi'(x_j)}, \tag{5.7}$$

$$\frac{w(x_j)p(x_j)u'(x_j)}{\phi'(x_j)} \simeq -\sum_{k=-M}^{N}\left\{\frac{\delta_{jk}^{(1)}}{h}(pw)(x_k)\right.\tag{5.8}$$

$$\left.+\delta_{jk}^{(0)}\left(\left(\frac{(pw)'}{\phi'}\right)(x_k)\right)\right\}u_k,$$

and

$$-\frac{w(x_j)u''(x_j)}{\phi'(x_j)} \simeq -\sum_{k=-M}^{N}\left\{\frac{\delta_{jk}^{(2)}}{h^2}(\phi'w)(x_k)\right.\tag{5.9}$$

$$\left.+\frac{\delta_{jk}^{(1)}}{h}\left(\frac{\phi''}{\phi'}w + 2w'\right)(x_k) - h\left(\frac{w''}{\phi'}\right)(x_k)\right\}u_k.$$

Finally, multiplying (5.2) by the factor $w(x_j)/\phi'(x_j)$ and replacing the terms of the resulting equation by (5.7)–(5.9) leads directly to (4.36). Thus the resulting system is given by

$$\left\{ \mathcal{A}(w) - \frac{1}{h} I^{(1)} D\left(\frac{p}{\phi'}\right) - D\left(\frac{(pw)' - qw}{(\phi')^2 w}\right) \right\} D(\phi'w)\vec{u} \quad (5.10)$$

$$= D\left(\frac{w}{\phi'}\right) \vec{f}.$$

This is the same as the system (4.39). If $w(x) = (\phi'(x))^{-r}$ in (5.10) then the system reads

$$\left\{ \mathcal{A}\left(\frac{1}{(\phi')^r}\right) - \frac{1}{h} I^{(1)} D\left(\frac{p}{\phi'}\right) + D\left(\frac{-1}{(\phi')^{2-r}} \left[\left(\frac{p}{(\phi')^r}\right)'\right]\right) \right. $$

$$\left. + D\left(\frac{q}{(\phi')^2}\right) \right\} D\left(\frac{1}{(\phi')^{r-1}}\right) \vec{u} = D\left(\frac{1}{(\phi')^{r+1}}\right) \vec{f} \quad (5.11)$$

where the matrix

$$\mathcal{A}\left(\frac{1}{(\phi')^r}\right) \equiv \frac{-1}{h^2} I^{(2)} - \frac{1}{h} I^{(1)} D\left((1-2r)\frac{\phi''}{(\phi')^2}\right) \quad (5.12)$$

$$+ D\left(\frac{-1}{(\phi')^{2-r}} \left(\frac{1}{(\phi')^r}\right)''\right).$$

The selection of the weight $w(x) = 1/\phi'(x)$ or $w(x) = 1/\sqrt{\phi'(x)}$ yields the systems (4.43) or (4.61), respectively, which correspond to $r = 1/2$ and 1, respectively.

A somewhat circuitous, but advantageous, development for a sinc collocation procedure for the differential equation (5.1) is motivated by Theorem 2.15. In contrast to Theorem 3.17, differentiation of the sinc expansion of $u(x)$ on the entire real line is subject to far less restrictive assumptions than is differentiation of $u(x)$ on a subset of the real line (compare the hypotheses of Theorem 2.15 with those of Theorem 3.17). The circuitous part of the development then dictates a transformation of the dependent variable in (5.1) to the entire real line. The advantageous portion of the development will play a more prevalent role in the next section on eigenvalue estimates.

For l a nonnegative real number, define the change of variable

$$v(t) = ((\phi')^l u) \circ \psi(t) \qquad (5.13)$$
$$= (\phi'(\psi(t)))^l u(\psi(t))$$

where ϕ and ψ are inverses of each other, as in Definition 3.1. A calculation using the chain rule and $x = \psi(t)$ leads to the equalities

$$u(x) = (\phi'(x))^{-l} v(x)$$
$$\frac{du}{dx} = (\phi'(x))^{-l} \frac{dv}{dx} + (\phi'(x)^{-l})' v(x)$$
$$\frac{d^2 u}{dx^2} = (\phi'(x))^{-l} \frac{d^2 v}{dx^2} + 2\left((\phi'(x))^{-l}\right)' \frac{dv}{dx} + (\phi'(x)^{-l})'' v(x).$$

Multiply (5.1) by $(\phi'(x))^l$, substitute each of the preceding equalities, and collect like derivatives to arrive at the transformed differential equation

$$-\frac{d^2 v}{dx^2} + \left\{\frac{2l\phi''(x)}{\phi'(x)} + p(x)\right\} \frac{dv}{dx}$$
$$+ \left\{-(\phi'(x))^l \left((\phi'(x))^{-l}\right)'' - \frac{l\phi''(x)p(x)}{\phi'(x)} + q(x)\right\} v(x)$$
$$= (\phi'(x))^l f(x).$$

Writing
$$\frac{dv}{dx} = \frac{dt}{dx} \frac{dv}{dt} = \phi'(x) \frac{dv}{dt}$$

and
$$\frac{d^2 v}{dx^2} = \frac{d}{dx}\left(\phi'(x) \frac{dv}{dt}\right) = \phi''(x) \frac{dv}{dt} + (\phi'(x))^2 \frac{d^2 v}{dt^2}$$

for the x-derivatives of the variable v and dividing the entire equation by $(\phi'(x))^2$ yields the equation

$$-\frac{d^2 v}{dt^2} + \left\{(2l-1) \frac{\phi''(x)}{\phi'(x)^2} + \frac{p(x)}{\phi'(x)}\right\} \frac{dv}{dt}$$
$$+ \left\{-(\phi'(x))^{l-2} \left((\phi'(x))^{-l}\right)'' - l \frac{\phi''(x)p(x)}{(\phi'(x))^3} + \frac{q(x)}{(\phi'(x))^2}\right\} v$$
$$\qquad (5.14)$$
$$= (\phi'(x))^{l-2} f(x).$$

If one wishes to calculate the various coefficients in (5.14), then repeated use of the identities $\phi'(x) = (\psi'(t))^{-1}$ and $dt/dx = \phi'(x) = 1/\psi'(t)$ will, with patience, yield expressions for the coefficients in terms of the variable t. For example

$$\phi''(x) = \frac{d}{dx}\left(\frac{1}{\psi'(t)}\right) = \frac{dt}{dx}\frac{d}{dt}\left(\frac{1}{\psi'(t)}\right) = \frac{-\psi''(t)}{(\psi'(t))^3}.$$

However, the change of variable is simply a catalyst, and in the end it will be more convenient to have the coefficients in the variable of the given problem (5.1). Hence denote the transformed boundary value problem by

$$\begin{aligned} Lv(t) &\equiv -v''(t) + \xi_p(t)v'(t) + \gamma_q(t)v(t) \\ &= (\psi'(t))^{2-l} f(\psi(t)), \quad -\infty < t < \infty \quad (5.15) \\ \lim_{t \to \pm\infty} v(t) &= 0 \end{aligned}$$

where the boundary conditions follow from $\psi(-\infty) = a$, $\psi(\infty) = b$, and the homogeneous boundary conditions for u in (5.1). The coefficients of dv/dt and $v(t)$ in (5.14) are given (explicitly in x) by

$$\xi_p(t) = \xi_p(\phi(x)) \equiv (2l-1)\frac{\phi''(x)}{(\phi'(x))^2} + \frac{p(x)}{\phi'(x)} \quad (5.16)$$

and

$$\begin{aligned} \gamma_q(t) &= \gamma_q(\phi(x)) \\ &\equiv \frac{-1}{(\phi'(x))^{2-l}}\left(\frac{1}{(\phi'(x))^l}\right)'' - \frac{l\phi''(x)p(x)}{(\phi'(x))^3} + \frac{q(x)}{(\phi'(x))^2}. \end{aligned} \quad (5.17)$$

The approximate sinc expansion

$$C_m(v,h)(t) = \sum_{k=-M}^{N} v(kh)S(k,h)(t), \quad m = M + N + 1 \quad (5.18)$$

and its derivatives

$$\frac{d^n}{dt^n}C_m(v,h)(t) = \sum_{k=-M}^{N} v(kh)\frac{d^n}{dt^n}S(k,h)(t) \quad (5.19)$$

STEADY PROBLEMS

provide accurate approximations to the functions $v^{(n)}$ ($n = 0, 1, 2$) as long as v satisfies the hypotheses of Theorem 2.16. Using each of (5.18) and (5.19) in (5.15) and evaluating this result at $t = jh$, $-M \leq j \leq N$, yields the identity

$$LC_m(v, h)(jh) = (\psi'(jh))^{2-l} f(\psi(jh))$$

which, upon letting $j = -M, \ldots, N$, has the more compact representation

$$C(l)\vec{v} = D((\psi')^{2-l})\vec{f} \qquad (5.20)$$

where

$$C(l) = \left\{ \frac{-1}{h^2} I^{(2)} - \frac{1}{h} D(\xi_p) I^{(1)} + D(\gamma_q) \right\} \qquad (5.21)$$

and $\vec{v} = (v(-Mh), \ldots, v(Nh))^T$, $\vec{f} = (f(-Mh), \ldots, f(Nh))^T$. An important point to this equation (with respect to the collocation procedure in Section 5.2) is that the vector $\vec{v}$ is the exact solution of (5.15) evaluated at the nodes $t_j = jh$.

For the present purpose, assume that an approximate solution of (5.15) is defined by

$$v(t) \simeq v_m(t) = \sum_{k=-M}^{N} v_k S(k, h)(t) \qquad (5.22)$$

and define the collocation scheme for (5.15) by the equation

$$C(l)\vec{v}^a = D((\psi')^{2-l})\vec{f} \qquad (5.23)$$

where $\vec{v}^a = (v_{-M}, \ldots, v_0, \ldots, v_N)^T$ are the approximations for the coefficients $v(kh)$ in (5.18).

From the definition of the system in (5.23), there are collocation schemes available for each choice of l. The particular values $l = 0, 1/2$, and 1 in the expressions for ξ_p (5.16) and for γ_q (5.17) all have a sort of distinguished role in the development. For the remainder of this chapter, the notation

$$\xi_p(x) = \begin{cases} \frac{-\phi''(x)}{(\phi'(x))^2} + \frac{p(x)}{\phi'(x)}, & l = 0 \\ \frac{p(x)}{\phi'(x)}, & l = 1/2 \\ \frac{\phi''(x)}{(\phi'(x))^2} + \frac{p(x)}{\phi'(x)}, & l = 1 \end{cases}$$

and

$$\gamma_q(x) = \begin{cases} \frac{q(x)}{(\phi'(x))^2}, & l = 0 \\ \frac{-1}{(\phi'(x))^{3/2}} \left(\frac{1}{(\phi'(x))^{1/2}}\right)'' - \frac{\phi''(x)p(x)}{2(\phi'(x))^3} + \frac{q(x)}{(\phi'(x))^2}, & l = 1/2 \\ \frac{-1}{(\phi'(x))} \left(\frac{1}{(\phi'(x))}\right)'' - \frac{\phi''(x)p(x)}{(\phi'(x))^3} + \frac{q(x)}{(\phi'(x))^2}, & l = 1 \end{cases}$$

will be adopted. The three matrices in (5.21), corresponding to these choices of l, define via the system (5.23) the coefficients $\vec{u} = (u_{-M}, \ldots, u_0, \ldots, u_N)^T$ in an approximate solution to the boundary value problem (5.1). This determination comes from the discretization of the variable change in (5.13)

$$\vec{v} = D((\phi')^l)\vec{u}.$$

That is, solve the system

$$C(l)D((\phi')^l)\vec{u} = D\left(\frac{1}{(\phi')^{2-l}}\right)\vec{f} \qquad (5.24)$$

for the coefficients in $\vec{u}$ and define the approximate solution of (5.1) by

$$u_m(x) = \sum_{k=-M}^{N} u_k S(k, h) \circ \phi(x). \qquad (5.25)$$

At least formally, the development is valid for any l (its worth is a different question), but the choices $l = 0, 1/2$, and 1 do have particular merit.

There is a very familiar system housed in (5.24). Recalling the self-adjoint case of (5.1) ($p(x) \equiv 0$), the boundary value problem reads

$$-u''(x) + q(x)u(x) = f(x), \quad a < x < b$$
$$u(a) = u(b) = 0 \qquad (5.26)$$

which is precisely the problem considered in (4.59). That is, if $l = 1/2$, then $\xi_p = 0$ and γ_q is the familiar expression from Table 4.7

$$\gamma_q(x) = \frac{-1}{(\phi'(x))^{3/2}} \left(\frac{1}{(\phi'(x))^{1/2}}\right)'' + \frac{q(x)}{(\phi'(x))^2} \qquad (5.27)$$

STEADY PROBLEMS

so that this collocation scheme and the Sinc-Galerkin scheme in (4.61) are identical.

It is of some interest to note that for the more general problem (5.1) ($p \not\equiv 0$) the matrix in (5.21) takes the form for $l = 0$

$$C(0) = \left\{ \frac{-1}{h^2} I^2 - \frac{1}{h} D\left(\frac{-\phi''}{(\phi')^2} + \frac{p}{\phi'}\right) I^{(1)} + D\left(\frac{q}{(\phi')^2}\right) \right\} \quad (5.28)$$

and for $l = 1$

$$C(1) = \left\{ \frac{-1}{h^2} I^{(2)} - \frac{1}{h} D\left(\frac{\phi''}{(\phi')^2} + \frac{p}{\phi'}\right) I^{(1)} \right.$$

$$\left. + D\left(\frac{-1}{\phi'}\left(\frac{1}{\phi'}\right)'' - \frac{\phi'' p}{(\phi')^3} + \frac{q}{(\phi')^2}\right) \right\}. \quad (5.29)$$

Which of the systems (5.11) or (5.24), with ($l = 0$ or 1), is best suited to handle the numerical solution of (5.1) is perhaps not clear from their representations above. A comparison of (5.12) with $r = 1$ and both of (5.28) and (5.29) show that the three systems share a similar structure. The coefficient matrix in each of these systems is the sum of three terms. The trailing diagonal matrices in (5.12) and (5.29) are the same ($p = q \equiv 0$) so that perhaps these two methods are (more or less) equivalent. The former performed very well on a wide class of problems throughout Section 4.3, and the same performance would be expected of (5.29). This expectation is borne out in Example 5.1. On the other hand, the trailing diagonal matrix in $C(0)$ is independent of p. Due to this simplicity, perhaps (5.24), with $C(0)$ as the coefficient matrix, is the distinguished system among the three.

At least as striking is the manner of entry of the two matrices $D(\xi_p)$ and $I^{(1)}$ in (5.12) ($r = 1$) compared to (5.28) or (5.29). Upon subtracting (5.28) from (5.12), one is led to an inspection of the commutator $I^{(1)}D(\xi_p) - D(\xi_p)I^{(1)}$. Exercise 5.1 suggests that this commutator behaves like a diagonal matrix. Although the exact role played by this commutator as an algebraic connection between the Sinc-Galerkin schemes (5.11) and the sinc-collocation schemes (5.24) is not analytically established, the result in Exercise 5.1 argues for a foundation based on its representation as a diagonal matrix. Exercise 5.2 shows that, in a special case, this commutator is the zero matrix.

Although the collocation systems in (5.24) and the Galerkin systems in (5.11) have different developments, they represent in the case of the self-adjoint problem (5.26) ($r = 1/2$ in (5.11) and $l = 1/2$ in (5.24)), the confluence of disjunctive objectives—they are identical. The numerical results in Example 5.1 and Exercise 5.3 in conjunction with the circumspection of Exercise 5.1 support the notion that (5.28) and (5.11) ($r = 1$) are directly connected. The matrix in (5.29) is perhaps more appropriate for the nonhomogeneous problem.

A sinc-collocation procedure with the intent of discretizing (5.1) in a manner that does not involve any terms with a derivative of p is carried out in [16]. The resulting system is a sort of cross-breeding of (5.11), with $r = 1$, and (5.29) and performs as well as either. A nice development of the scheme in (5.24), which is based on Theorem 3.17, is found in [2]. This seemingly scattered array of schemes may appear bewildering, but there is a harmony in the confusion. It has been established numerically that they all perform well, if not equivalently, on (5.1). In the case of $r = l = 1/2$ in (5.11) and (5.24), as pointed out above, the schemes are identical.

If the focus is only on the problem (5.1), then it really does not matter which of the above schemes one selects. However, in the case of more complicated problems (partial differential equations), it is easy to envision a preference of one scheme over another, for example, with respect to ease of implementation. Independent of personal prejudice, there is comfort in knowing that the exponential accuracy of Theorem 4.7 is obtained for each of the available choices.

Example 5.1 The problem

$$-u''(x) - \frac{1}{6x} u'(x) + \frac{1}{x^2} u(x) = \frac{19}{6} \sqrt{x}, \quad 0 < x < 1$$
(5.30)
$$u(0) = u(1) = 0$$

was computed via the Sinc-Galerkin method in Example 4.5. From that example the discrete system for the weight function $w(x) = 1/\phi'(x)$ has the form

$$\left\{ \frac{-1}{h^2} I^{(2)} + \frac{1}{h} I^{(1)} D\left(\frac{11x-5}{6}\right) + D\left((1-x)\left(\frac{5x+6}{6}\right)\right) \right\} \vec{u}$$
$$= D\left(\frac{19}{6} x^{5/2}(1-x)^2\right) \vec{1}$$

STEADY PROBLEMS

The collocation form of the discrete system is, from (5.29) (for $l = 1$),

$$\left\{\frac{-1}{h^2} I^{(2)} - \frac{1}{h} D\left(\frac{13x-7}{6}\right) I^{(1)} + D\left((1-x)\left(\frac{8x+5}{6}\right)\right)\right\} \vec{u}$$
$$= D\left(\frac{19}{6} x^{3/2}(1-x)\right) \vec{1}.$$

Since the true solution is $u(x) = x^{3/2}(1-x)$, one can select $\alpha = 3/2$ and $\beta = 1$. See the discussion in Section 4.5 for the selections of α and β based on series solutions. Recalling (4.35) and (4.29), the selections $N = 3M/2$ and $h = \pi/\sqrt{3M}$ ($d = \pi/2$) are made in Table 5.1. For purposes of comparison, the Sinc-Galerkin error on the sinc gridpoints (reported as $\|E_S^G(h)\|$) is repeated from Table 4.2. The sinc-collocation error (using the matrix (5.29)) on the sinc gridpoints is reported under the column headed $\|E_S^C(h)\|$. If the Sinc-Galerkin method appears to be the winner in this example, then the odds are evened out in Exercise 5.3, where the calculation of $\|E_S^C(h)\|$ using the matrix (5.28) yields numerically identical results when compared with $\|E_S^G(h)\|$. This is the conjectured sinc-collocation equivalence to the Sinc-Galerkin system (5.11) with $r = 1$.

M	N	$h = \pi/\sqrt{3M}$	$\|E_S^G(h)\|$	$\|E_S^C(h)\|$
4	6	0.907	$.208 \times 10^{-2}$	$.195 \times 10^{-2}$
8	12	0.641	$.262 \times 10^{-3}$	$.266 \times 10^{-3}$
16	24	0.453	$.127 \times 10^{-4}$	$.293 \times 10^{-4}$
32	48	0.321	$.157 \times 10^{-6}$	$.628 \times 10^{-5}$

Table 5.1 Comparison of the errors in the Sinc-Galerkin ($r = 1$) and the sinc-collocation ($\ell = 1$) solution of (5.30).

5.2 Sturm–Liouville Problems

There are various forms by which one can define a Sturm–Liouville problem, each of which has its particular merits depending on the problem at hand. Throughout this section a *Sturm–Liouville problem* is a system consisting of a differential equation written in the *normal form*

$$-\frac{d^2u}{dx^2} + q(x)u(x) = \lambda\rho(x)u(x), \quad a < x < b \qquad (5.31)$$

and the separated endpoint conditions

$$a_0 u(a) - a_1 u'(a) = 0$$
$$b_0 u(b) + b_1 u'(b) = 0. \qquad (5.32)$$

The system is called *regular* if the interval is finite, the function ρ is positive on $[a, b]$, and q and ρ are continuous. If any of these conditions is not met, the Sturm–Liouville system is called *singular*. For example, *Bessel's differential equation of order ν*, $\nu \geq 1$, is given by

$$-\frac{d^2u}{dx^2} + \left(\frac{4\nu^2 - 1}{4x^2}\right)u(x) = \lambda u(x), \quad 0 < x < 1$$
$$u(0) = u(1) = 0. \qquad (5.33)$$

This problem is singular because the function $q(x) = (4\nu^2 - 1)/(4x^2)$ is unbounded as $x \to 0^+$.

If the interval (a, b) in (5.31) is semi-infinite or infinite, the system is also called singular. In this event there is a corresponding change in the form of the end conditions in (5.32). For example, the *Hermite differential equation* takes the form

$$-\frac{d^2u}{dx^2} + \frac{x^2}{4}u(x) = \lambda u(x), \quad -\infty < x < \infty$$
$$\lim_{x \to \pm\infty} u(x) = 0 \qquad (5.34)$$

so that $\rho(x) \equiv 1$ and $q(x) = x^2/4$ in (5.31). The limit condition in (5.34) replaces the end conditions at a and b in (5.32). The system in (5.34) is singular due to the infinite interval.

The self-adjoint form of the Sturm–Liouville problem (5.31) is defined by

$$\frac{-d}{dx}\left(c(x)\frac{dy}{dx}\right) + q(x)y(x) = \lambda \rho(x) y(x), \quad a < x < b \qquad (5.35)$$

where for a regular problem it is assumed, in addition to the above, that the coefficient $c(x)$ is in $C^1[a,b]$ and positive on $[a,b]$. For example, the Bessel differential equation via the change of variable

$$y(x) = x^{-1/2} u(x)$$

is transformed from (5.33) into the form

$$-\frac{d}{dx}\left(x\frac{dy}{dx}\right) + \frac{\nu^2}{x} y(x) = \lambda x y(x), \quad 0 < x < 1$$

$$u(0) = u(1) = 0.$$

For ν an integer the solutions are the (analytic) Bessel functions of the first kind $J_\nu(x)$, whereas the solutions of (5.33) have an algebraic singularity at $x = 0$. The eigenvalues are the same for either form of the system. Whether one chooses to worry about the eigenfunction singularity in (5.34) is somewhat dependent on available computational algorithms.

Motivated by the sort of application stemming from the radial Schrödinger equation, the form in (5.31) with $a_1 = b_1 = 0$ in (5.32) is adhered to throughout this section. The case of the Neumann or radiation conditions in (5.32) is a bit more delicate and is the subject of the work in [10]. As far as the assumed normal form assumption goes, the transformation $y(x) = (c(x))^{-1/2} u(x)$ converts (5.35) into the normal form (5.31). A sinc discretization for the self-adjoint form in (5.35) will be developed in the next section, but with a very different goal in mind.

The theorem below gives the appropriate setting for the exponential convergence of the eigenvalue approximation to a true eigenvalue of the Sturm–Liouville problem

$$-\frac{d^2u}{dx^2} + q(x)u(x) = \lambda\rho(x)u(x), \quad a < x < b, \tag{5.36}$$

$$u(a) = u(b) = 0.$$

If the change of variable

$$v(t) = \left(\sqrt{\phi'}\, u\right) \circ \psi(t) \tag{5.37}$$

is made in (5.36), then with $l = 1/2$ in (5.15) the transformed problem is

$$-v''(t) + \gamma_q(t)v(t) = \lambda\rho(\psi(t))(\psi'(t))^2 v(t), \quad -\infty < t < \infty, \tag{5.38}$$

$$\lim_{t \to \pm\infty}(v(t)) = 0,$$

where $\gamma_q(t)$ is defined in (5.17). From (5.20) and (5.21), this gives the collocation scheme

$$C\left(\frac{1}{2}\right)\vec{v} = \lambda D\left(\rho(\psi) \cdot (\psi')^2\right)\vec{v} \tag{5.39}$$

for the differential equation (5.38) on $-\infty < t < \infty$. The notation $C(1/2)$ is consistent with (5.21) as long as it is recalled that for $p(x) = 0$ the coefficient ξ_p of the matrix $I^{(1)}$ is zero. That is, for $p = 0$ the matrix is the same as the matrix in (5.21) with $l = 1/2$. It is repeated in (5.40) below. The vector of point evaluations

$$\vec{v} = (v(-Mh), \ldots, v(0), \ldots, v(Nh))^T$$

are the coefficients in the sinc expansion of

$$C_m(v, h)(t) = \sum_{k=-M}^{N} v(kh)S(k, h)(t), \quad m = M + N + 1.$$

It is important in the proof of the theorem (the equality in (5.45)) below that the coefficients in $C_m(v, h)$ are the true solution of the eigenvalue problem at the nodes (as opposed to the approximate coefficients in (5.22)).

STEADY PROBLEMS

The eigenvalues of the transformed problem (5.38) and the original problem (5.36) are identical. Thus in practice the transformation need not be done. It merely facilitates the proof of the eigenvalue error bound. The generalized eigenvalue problem for the original Sturm–Liouville problem (5.36) is then given by

$$C\left(\frac{1}{2}\right)\vec{z} = \left\{\frac{-1}{h^2}I^{(2)} + D\left(\frac{-1}{(\phi')^{3/2}}\left(\frac{1}{(\phi')^{1/2}}\right)'' + \frac{q}{(\phi')^2}\right)\right\}\vec{z} \quad (5.40)$$

$$= \mu D\left(\frac{\rho}{(\phi')^2}\right)\vec{z}$$

or

$$C\left(\frac{1}{2}\right)\vec{z} = \mu \mathcal{D}^2 \vec{z}$$

where

$$\mathcal{D} = D\left(\sqrt{\rho}/(\phi')\right) \quad \vec{z} = D\left(\sqrt{\phi'}\right)\vec{u}.$$

The complete statement of the eigenvalue error estimate is in the following theorem [8].

Theorem 5.2 *Let (λ_0, u_0) be an eigenpair of the Sturm–Liouville problem (5.36). Assume that $\sqrt{\phi'}\, u_0 \circ \psi \in B(D_S)$ and there are positive constants α, β, and C so that*

$$\left|\sqrt{\phi'(x)}\, u_0(x)\right| \leq C \begin{cases} \exp(-\alpha|\phi(x)|), & x \in \Gamma_a \\ \exp(-\beta|\phi(x)|), & x \in \Gamma_b \end{cases} \quad (5.41)$$

where

$$\Gamma_a = \{\psi(t) : t \in (-\infty, 0]\}, \quad \Gamma_b = \{\psi(t) : t \in (0, \infty)\}.$$

If there is a constant $\delta > 0$ so that $\gamma_q(x) \geq \delta^{-1}$, where

$$\gamma_q(x) = -(\phi'(x))^{-3/2}\left((\phi'(x))^{-1/2}\right)'' + (\phi'(x))^{-2}q(x)$$

and the selections

$$N = \left[\left|\frac{\alpha}{\beta}M + 1\right|\right], \quad h = \left(\frac{\pi d}{\alpha M}\right)^{1/2}$$

are made, then there is an eigenvalue μ_p of the generalized eigenvalue problem (5.40) satisfying

$$|\mu_p - \lambda_0| \leq K(\delta\lambda_0)^{1/2}M^{3/2}\exp\left(-(\pi d\alpha M)^{1/2}\right). \quad (5.42)$$

Proof Since
$$\gamma_q(x) \geq \delta^{-1} > 0, \quad x \in (a,b) \tag{5.43}$$
the symmetric matrix $C(1/2)$ in (5.40) is positive definite. Assume that $(\lambda_0, v_0(t))$ is an eigenpair of (5.38), where v_0 is normalized by
$$\int_{-\infty}^{\infty} v_0^2(t)\rho(\psi(t))[\psi'(t)]^2 dt = 1. \tag{5.44}$$
Note that with the change of variable defined by (5.37) the equality in (5.44) is equivalent to the assumption that
$$\int_a^b u_0^2(x)\rho(x)dx = 1.$$
Upon substituting v_0 in (5.37) and evaluating at $t = jh$ ($-M \leq j \leq N$), it follows that
$$\vec{L}v_0 = \lambda_0 D(\rho(\psi')^2)\vec{v}_0.$$
Subtracting this from (5.39) yields the equalities
$$\vec{\Delta v_0} = C(1/2)\vec{v}_0 - \vec{L}v_0$$
$$= \left(C(1/2) - \lambda_0 \mathcal{D}^2\right) \vec{v}_0 \tag{5.45}$$
where
$$\mathcal{D} = D\left(\sqrt{\rho}\psi'\right).$$
Since $C(1/2)$ and $\mathcal{D}^2$ are positive definite, there are eigenvectors $\vec{z}_i$ and positive eigenvalues $\mu_i \leq \mu_j$ ($-M \leq i < j \leq N$) (Theorem A.13) so that
$$Z^T C(1/2) Z = \begin{bmatrix} \mu_{-M} & & & & \\ & \ddots & & & \\ & & \mu_0 & & \\ & & & \ddots & \\ & & & & \mu_N \end{bmatrix}, \tag{5.46}$$
$$Z^T \mathcal{D}^2 Z = I$$
and
$$C(1/2)\vec{z}_i = \mu_i \mathcal{D}^2 \vec{z}_i, \quad i = -M, \ldots, N. \tag{5.47}$$

The independence of the $\{\vec{z}_i\}_{i=-M}^N$ implies that there are constants β_i so that

$$\vec{v}_0 = \sum_{i=-M}^{N} \beta_i \vec{z}_i. \qquad (5.48)$$

Substituting (5.48) in the right-hand side of (5.45) yields the equality

$$\vec{\Delta v_0} = \sum_{i=-M}^{N} \beta_i(\mu_i - \lambda_0)\mathcal{D}^2 \vec{z}_i.$$

Taking the inner product of the eigenvector $\vec{z}_p$ with each side of this equation shows that

$$\vec{z}_p^T \vec{\Delta v_0} = \beta_p(\mu_p - \lambda_0), \quad p = -M, \ldots, N. \qquad (5.49)$$

Applying $\mathcal{D}^2$ to both sides of (5.48) and taking the inner product of this result with $\vec{v}_0$ leads, upon using (5.47), to the identity

$$\|\mathcal{D}\vec{v}_0\|_2^2 = \sum_{i=-M}^{N} \beta_i^2 \leq \beta_p^2(M+N+1) \qquad (5.50)$$

where

$$|\beta_p| \equiv \max_{-M \leq i \leq N}\{|\beta_i|\}. \qquad (5.51)$$

Assume that the eigenfunction v_0 of (5.38) is an element of $B(D_S)$ and that v_0 satisfies the growth condition in (2.35). Application of the trapezoidal quadrature rule (2.51) to the left-hand side of (5.44) shows that

$$\begin{aligned} 1 &= h \sum_{j=-M}^{N} v_0^2(jh)\rho(\psi(jh))[\psi'(jh)]^2 + \eta_m(h) \\ &= h\|\mathcal{D}\vec{v}_0\|_2^2 + \eta_m(h). \end{aligned}$$

The assumptions on v_0 then guarantee that $|\eta_m(h)/h| \to 0$ as $m \to \infty$ so that, for sufficiently large m, $|\eta_m(h)/h| \leq 1/(2h)$. Hence $\|\mathcal{D}\vec{v}_0\|_2^2 \geq 1/(2h)$ which, when combined with (5.50), yields

$$|\beta_p| \geq (2mh)^{-1/2}. \qquad (5.52)$$

The inequality (5.43) shows that (letting $\alpha_{-M} \equiv \min\limits_{\alpha_j \in \sigma(C(1/2))}\{\alpha_j\}$)

$$\begin{aligned}\delta^{-1} &\leq \alpha_{-M} \\ &= \min_{\vec{x}^T\vec{x}}\left\{\frac{\vec{x}^T C(1/2)\vec{x}}{\vec{x}^T\vec{x}}\right\} \\ &\leq \vec{z}_j^T C(1/2)\vec{z}_j / \|\vec{z}_j\|_2^2.\end{aligned}$$

The equality in (5.47) shows that $\vec{z}_j^T C(1/2)\vec{z}_j = \mu_j$, which, when substituted into the above equation, leads to the estimate

$$\|\vec{z}_j\|_2^2 \leq \delta \mu_j. \tag{5.53}$$

Now let p denote the index implicitly defined in (5.51) and assume

$$|\mu_p - \lambda_0| \leq \lambda_0. \tag{5.54}$$

In what follows, the latter condition will be shown to hold for all sufficiently large m. If θ_p is the angle between the vectors $\vec{z}_p$ and $\vec{\Delta v_0}$, then taking absolute values in (5.49) and expanding the inner product on the left-hand side yields

$$|\mu_p - \lambda_0| = \frac{\|\vec{z}_p\|_2 \|\vec{\Delta v_0}\|_2 |\cos(\theta_p)|}{|\beta_p|}. \tag{5.55}$$

The inequality in (5.54) shows that

$$\mu_p = \mu_p - \lambda_0 + \lambda_0 \leq |\mu_p - \lambda_0| + \lambda_0 \leq 2\lambda_0$$

which when substituted in (5.53) gives

$$\|\vec{z}_p\|_2^2 \leq 2\lambda_0 \delta. \tag{5.56}$$

Replacing $\|\vec{z}_p\|^2$ and $|\beta_p|$ in (5.55) by the right-hand sides of (5.56) and (5.52), respectively, leads to

$$|\mu_p - \lambda_0| \leq 2\left\{(mh)^{1/2}|\cos(\theta_p)|\right\}\sqrt{\delta\lambda_0}\,\|\vec{\Delta v_0}\|_2. \tag{5.57}$$

A more convenient form for the error in (5.57) may be obtained by a short computation using (5.45) and the interpolatory property

STEADY PROBLEMS

$S(k,h)(jh) = \delta_{jk}$. This leads to

$$|\Delta v_0(jh)| = |LC_m(v_0,h)(jh) - Lv_0(jh)|$$

$$= \left|\frac{d^2}{dt^2} C_m(v_0,h)(jh) - \frac{d^2}{dt^2} v_0(jh)\right|.$$

Combining this identity with (2.44) and the inequality $\|\vec{\Delta v_0}\|_2 \leq \sqrt{m}\|\vec{\Delta v_0}\|_\infty$ shows that

$$\|\vec{\Delta v_0}\|_2 \leq KM^{5/4} \exp\left(-(\pi d\alpha M)^{1/2}\right). \tag{5.58}$$

Finally, the substitutions $h = (\pi d/(\alpha M))^{1/2}$, $|\cos\theta_p| \leq 1$, and the right-hand side of (5.54) for $\|\vec{\Delta v_0}\|_2$ in (5.57) yield the error estimate

$$|\mu_p - \lambda_0| \leq K\sqrt{\delta\lambda_0}\, M^{3/2} \exp\left(-(\pi d\alpha M)^{1/2}\right). \tag{5.59}$$

Note that in the complementary case to (5.54),

$$|\mu_p - \lambda_0| > \lambda_0. \tag{5.60}$$

The inequality (5.59) is then replaced by

$$\mu_p \leq 2|\mu_p - \lambda_0|.$$

In this case the right-hand side of (5.56) is replaced by $2\delta|\mu_p - \lambda_0|$. Hence (5.58) is the same as listed if $\lambda_0^{1/2}$ on the right-hand side of (5.59) is replaced by $|\mu_p - \lambda_0|^{1/2}$. With this replacement, (5.59) remains valid. Hence $|\mu_p - \lambda_0| \to 0$ as $m \to \infty$. While (5.60) may be valid for some values of m, as m increases (5.54) takes over.

∎

The appropriate conformal map for (5.36) is determined by the interval (a,b) and the asymptotic behavior of the eigenfunctions. If the interval is finite, the map used is $\phi(x) = ln((x-a)/(b-x))$. For the semi-infinite interval $(0,\infty)$, the two available maps are given by $\phi(x) = ln(x)$ or $ln(\sinh(x))$. The criteria discussed at the close of Section 4.4 are applicable here. In many Sturm–Liouville problems both of the mappings work quite well and, in this case, the map $\phi(x) = ln(x)$ is perhaps preferable due to the simpler discrete system. These maps and the terms of the discrete system (5.40) are given in Table 4.7. Example 5.3 illustrates a situation where both maps are applicable but, if one is willing to augment the selection with a little analytic work, noticeably improved calculations are achieved with the map $ln(\sinh(x))$. The maps $\phi(x) = x$ and $\sinh^{-1}(x)$ in Table 4.7 are useful for problems on the real line. The identity map yields a direct sinc discretization of problems on the entire real line (the trapezoidal rule) such as Hermite's differential equation in (5.34). Example 5.4 implements this trapezoidal rule on the Hermite differential equation and explores some eigenfunction parity via matrix splitting. The inverse sinh map is often used with the goal of accelerating the convergence of a method. The situation here is very similar to that in quadrature. As a rule of thumb, it is recommended only for the more stubborn problems.

The assumption that $|\gamma_q(x)| \geq \delta^{-1}$, upon an inspection of Table 4.7, shows that δ may be replaced by 4 in (5.59) in the case of the maps $\phi(x) = ln((x-a)/(b-x))$ and $\phi(x) = ln(x)$. If, in the case of the map $ln(\sinh(x))$, $\liminf_{x \to \infty} q(x) \neq 0$, then there exists a δ so that (5.43) is satisfied. However whether or not δ is bounded away from zero is problem dependent.

Example 5.4 The harmonic oscillator is governed by

$$-u''(x) + (x^2 + \gamma x^{-2})u(x) = \lambda u(x), \quad 0 < x < \infty$$
(5.61)
$$u(0) = \lim_{x \to \infty} u(x) = 0.$$

This is a singular Sturm–Liouville problem due to both the infinite interval $(0, \infty)$ and the singularity in $q(x) = x^2 + \gamma x^{-2}$ at $x = 0$ (for $\gamma \neq 0$). The eigenfunctions may have singularities depending on the value of γ. If $\gamma = 0$, the equation has analytic coefficients and the eigenfunctions are a subset of the Hermite functions, which are entire (see Example 5.4). If $\gamma = 2$, (5.61) is the radial Schrödinger equation with harmonic oscillator potential. In this case q, is singular but the eigenfunctions are again entire. However, if $\gamma = 3/4$, the eigenfunctions have a branch singularity at $x = 0$. This case is treated here.

To justify the remarks of the previous paragraph, one can compute the eigenfunctions of (5.61) which, for arbitrary γ, are given by

$$u_\lambda(x) = x^{r(\gamma)} y_\lambda(x) e^{-x^2/2} \qquad (5.62)$$

where

$$r(\gamma) = (1 + \sqrt{4\gamma + 1})/2$$

and the $y_\lambda(x)$ satisfy

$$-xy''(x) + 2(x^2 - r(\gamma))y'(x) + (2r(\gamma) + 1)xy(x) = \lambda(\gamma)xy(x).$$

The associated eigenvalues $\lambda(\gamma)$ are given by

$$\lambda_j(\gamma) = 2r(\gamma) + 1 + 4(j - 1), \quad j = 1, 2, 3, \ldots.$$

Turning to the case $\gamma = 3/4$, it follows from (5.62) that the eigensolutions have a branch singularity, since $r(3/4) = 3/2$. Now one has the option of using either $\phi(x) = \ell n(x)$ or $\ell n(\sinh(x))$ (both satisfy the bound in (5.41)). The work in [5] (a close inspection of the bound on $N(u, D_W)$ in (3.4)) shows that d is bounded above by $\pi/4$, whereas, due to the boundedness in the imaginary direction of the domain D_B, one may select $d = \pi/2$ for the $\ell n(\sinh(x))$ map. A comparison of the results below with the results in [14] shows a difference in accuracy that is predicted by this discrepancy in the available selections of d. The results in [14] are certainly not bad, but the results below are the more accurate of the two for the same system size. This same situation is illustrated again in Exercise 5.5.

The discrete system for (5.61) takes the form

$$\left[-\frac{1}{h^2} I^{(2)} + D\left(\frac{4\cosh^2(x) - 3}{4\cosh^4(x)} + (x^2 + \frac{\gamma}{x^2})\tanh^2(x)\right)\right] \vec{z}$$

$$= \mu D(\tanh^2(x))\vec{z}.$$

The errors between the computed eigenvalues μ_p and the true eigenvalues λ_p of (5.61) are reported in Table 5.2 as

$$|E(\lambda_p)| = |\lambda_p - \mu_p| = |4p - \mu_p|.$$

The parameters used in the discrete system are $\alpha = r(3/4) - 1/2 = 1$, $\beta = 1$, $d = \pi/2$, $h = \pi/\sqrt{2M}$, and $N = M$.

$M=N$	$h=\pi/\sqrt{2M}$	$|E(\lambda_1)|$	$|E(\lambda_2)|$	$|E(\lambda_3)|$
4	1.111	$.499 \times 10^{-1}$		
8	0.785	$.222 \times 10^{-2}$	$.106 \times 10^{-2}$	
16	0.555	$.926 \times 10^{-5}$	$.183 \times 10^{-3}$	$.680 \times 10^{-3}$

Table 5.2 Errors in the eigenvalues of (5.61) ($\gamma = 3/4$) via the sinc-collocation method with the map $\phi(x) = \ell n(\sinh(x))$.

The problem (5.61) with $\gamma = 2$ was computed in [5], and a comparison of the results found there with the results above shows the former are about a digit more accurate. This is because of the larger value of α that corresponds to $\gamma = 2$ ($r(2) = 2$ implies $\alpha = 3/2$). ∎

Example 5.4 The Hermite differential equation is defined by

$$-u''(x) + \frac{x^2}{4}u(x) = \lambda u(x), \quad -\infty < x < \infty$$

$$\lim_{x \to \pm\infty} u(x) = 0.$$
(5.63)

The computation of the eigenvalues of the Hermite equation, as well as of other Sturm–Liouville systems exhibiting even and odd parity, is sometimes dealt with by considering the even and odd eigenfunctions separately. To calculate the eigenvalues associated with the odd eigenfunctions, it is necessary to solve only

$$-u''(x) + \frac{x^2}{4}u(x) = \lambda u(x), \quad 0 < x < \infty$$

$$u(0) = 0, \quad \lim_{x \to \infty} u(x) = 0.$$
(5.64)

Similarly, the eigenvalues associated with the even eigenfunctions may be calculated from

$$-u''(x) + \frac{x^2}{4}u(x) = \lambda u(x), \quad 0 < x < \infty$$

$$u(0) = c, \quad \lim_{x \to \infty} u(x) = 0$$

where the eigenfunctions are normalized by $u(0) = c \neq 0$.

This appears to be a considerable amount of work, which could be avoided by merely dealing directly with the Hermite equation (5.63). This problem, as well as others that have symmetric potential and weight functions, may be treated as just described or the parity separation may be accomplished through a factorization of the discrete system. The latter is the method described in this example [11]. Due to the Toeplitz structure of the system (5.40), the sinc-collocation method applied to (5.63) admits a splitting in the discrete system analogous to the even and odd eigenfunction splitting of (5.63). To see this, set $m = 2M + 1$ ($M = N$) and write the coefficient matrix in (5.40) in the form

$$\mathcal{C}(1/2) = -\frac{1}{h^2} I^{(2)} + \kappa I + D(r)$$

where, from Table 4.7, $\kappa = 0$ or $1/4$ and $r(-x) = r(x)$ in the diagonal matrix $D(r)$. Here $r(x) = q(x)/(\phi'(x))^2$. This matrix admits the representation

$$\mathcal{C}\left(\frac{1}{2}\right) = \frac{-1}{h^2} \begin{bmatrix} I_M^{(2)} & \vec{c}_{M \times 1} & L_M^T \\ (\vec{c}_{M \times 1})^T & \pi^2/3 & (\vec{c}_{M \times 1})^T J \\ L_M & J\vec{c}_{M \times 1} & I_M^{(2)} \end{bmatrix}$$

$$+ \kappa I_m + \begin{bmatrix} D_-(r) & & \\ & r(0) & \\ & & D_+(r) \end{bmatrix}$$

where the subscripts denote the block sizes, $I_M^{(2)}$ is the $M \times M$ matrix $I^{(2)}$, $\vec{c}_{M \times 1}$ consists of the first M elements of the $(M+1)$-st column of $I_m^{(2)}$, and L_M is the $M \times M$ portion of $I_m^{(2)}$ in its lower left-hand corner. The $M \times M$ matrices

$$D_{\mp}(r) = \text{diag}(r(x_{\mp k}))$$

are connected by the $M \times M$ counteridentity (see (A.11))

$$J = \begin{bmatrix} & & & 1 \\ & & \cdot & \\ & \cdot & & \\ 1 & & & \end{bmatrix}$$

via

$$D_-(r) = JD_+(r)J \qquad (5.65)$$

since r is an even function and $x_{-k} = -x_k$.

The $m \times m$ orthogonal matrix

$$Q = \frac{1}{\sqrt{2}} \begin{bmatrix} I_M & \vec{0}_{M \times 1} & I_M \\ (\vec{0}_{M \times 1})^T & \sqrt{2} & (\vec{0}_{M \times 1})^T \\ -J_M & \vec{0}_{M \times 1} & J_M \end{bmatrix}$$

block diagonalizes $\mathcal{C}(1/2)$, since

$$Q^* \mathcal{C}(1/2) Q = \frac{-1}{h^2} \begin{bmatrix} I_M^{(2)} - JL_M & \vec{0}_{M \times 1} & 0_M \\ (\vec{0}_{M \times 1})^T & \pi^2/3 & \sqrt{2}(\vec{c}_{M \times 1})^T \\ 0_M & \sqrt{2}\,\vec{c}_{M \times 1} & I_M^{(2)} + JL_M \end{bmatrix}$$

$$+ \kappa I_m + \begin{bmatrix} D_-(r) & \vec{0}_{M \times 1} & 0_M \\ (\vec{0}_{M \times 1})^T & r(0) & (\vec{0}_{M \times 1})^T \\ 0_M & \vec{0}_{M \times 1} & D_+(r) \end{bmatrix}. \qquad (5.66)$$

Hence the spectrum of $\mathcal{C}(1/2)$ is the same as the union of the spectrums of the two matrices

$$C^- = \frac{-1}{h^2}\left(I_M^{(2)} - JL_M\right) + \kappa I_M + D_-(r)$$

and

$$C^+ = \frac{-1}{h^2} \begin{bmatrix} \pi^2/3 & \sqrt{2}(\vec{c}_{M \times 1})^T \\ \sqrt{2}\,\vec{c}_{M \times 1} & I_M^{(2)} + JL_M \end{bmatrix}$$

$$+ \kappa I_{M+1} + \begin{bmatrix} r(0) & (\vec{0}_{M \times 1})^T \\ (\vec{0}_{M \times 1}) & D_+(r) \end{bmatrix}.$$

STEADY PROBLEMS   187

In the case where ρ in (5.31) is symmetric (it is identically 1 in (5.63)) the eigenvalues of $C^{\mp}$ are the approximate eigenvalues of the continuous problem. In the case where ρ is nonconstant but symmetric, this follows from (5.65) with r replaced by ρ.

This splitting would be aesthetically complete if one could assert that the spectrums of C^- and C^+ are interlaced. Such a result implies, for example, that the approximate smallest eigenvalue of the continuous problem is contained in the spectrum of C^+. Whereas this has been verified in numerical computation, an analytic statement along these lines does not appear to follow from an elementary argument. Techniques such as Cauchy's Interlace Theorem A.16 give only crude estimates on eigenvalue intermixing due to the large size of C^+ (or C^-) relative to the size of $C(1/2)$. This splitting is not limited to the Hermite problem. The first matrix on the right-hand side of (5.66) is independent of the problem. This statement remains true with the addition of κI in the case of problems on (a,b) and $(0,\infty)$ ($\phi(x) = \ell n(x)$), since κ is then $1/4$. The final matrix in (5.66) is obtained for the finite interval $(-a, a)$ for symmetric q and ρ, since (5.65) follows from $x_k = \frac{a(e^{kh}-1)}{e^{kh}+1} = -x_{-k}$. The numerical results for the Hermite differential equation (5.63) are given in Table 5.3, where the error in the p-th eigenvalue $\lambda_p = 1/2, 3/2,$ and $5/2$ is denoted by $|E(\lambda_p)|$.

M	$h = \pi/\sqrt{2M}$	$|E(\lambda_1)|$	$|E(\lambda_2)|$	$|E(\lambda_3)|$
4	1.111	$.115 \times 10^{-5}$	$.592 \times 10^{-4}$	$.344 \times 10^{-3}$
8	0.785	$.145 \times 10^{-9}$	$.675 \times 10^{-8}$	$.151 \times 10^{-6}$
12	0.641	$.139 \times 10^{-14}$	$.646 \times 10^{-12}$	$.206 \times 10^{-10}$

Table 5.3 Eigenvalue errors of (5.63) from the reduced or split system.

Using the matrix C^+, the smallest eigenvalue of (5.63) is computed with an accuracy of 5, 9, and 14 digits with systems of size 5,

9, and 13, respectively. This is the third column (labeled $|E(\lambda_1)|$) of Table 5.3. For this example the spectrums of C^+ and C^- were compared and their spectrums do indeed interlace. The surprising accuracy is explained upon recalling the result in Exercise 2.1 in conjunction with the fact that the Hermite functions are entire functions. More specifically, this sort of "super" convergence is justified on a class of eigenproblems of which the Hermite problem is probably the most familiar example. The numerical results for the anharmonic oscillator problem ($q(x) = x^2 + \gamma x^4$ and $\rho(x) = 1$ in (5.36)) yield similar impressive results [9].

∎

5.3 Discretization of a Self-Adjoint Form

This section addresses a discretization of the operator

$$L(c)u(x) \equiv -\frac{d}{dx}\left(c(x)\frac{du}{dx}\right) \tag{5.67}$$

where the focus is twofold. One purpose is to isolate the coefficient c in the discrete system, and the other is to remove all differentiations from c in the resulting discrete system. There are a number of approaches [18] one could take for the discretization of this self-adjoint form based on the methodologies in both Section 4.3 and Section 5.1, but the procedure of the present section emphasizes a little more directly the interpolatory nature of sinc approximation. If the goal is to solve for the solution u, these previous procedures work fine as they stand. For example, if $l = 1/2$ in (5.21), then a discretization of (5.67), subject to homogeneous Dirichlet boundary conditions, takes the form

$$\left\{\frac{-1}{h^2}D(c)I^{(2)} + \frac{1}{h}D\left(\frac{c'}{\phi'}\right)I^{(1)} + D\left(\frac{1}{4}c + \frac{\phi''c'}{(\phi')^3}\right)\right\}D(\sqrt{\phi'})\vec{u}$$

$$= D\left(\frac{1}{(\phi')^{3/2}}\right)\vec{f} \tag{5.68}$$

and may be directly solved for $\vec{u}$. However, if the goal is to solve for the coefficient c, then the above has the drawback of requiring the

STEADY PROBLEMS 189

derivative of c in the discrete system. It is also a little unclear how to solve the above system for c given that one knows the solution u or samples of the solution.

The model problem considered here seeks to find c in

$$L(c)u(x) \equiv -\frac{d}{dx}\left(c(x)\frac{du}{dx}\right) = f(x), \quad 0 < x < 1$$

$$u(0) = u(1) = 0 \tag{5.69}$$

given sample data of the true solution u. Independent of the manner of discretization of the expression in (5.69) the problem, as stated, has infinitely many solutions. There are conditions one can impose on the coefficient c to single out a unique solution, but to go into these matters here would lead far afield. The excellent text [1] includes a very readable account of conditions to impose on c in order that the coefficient can be determined. Even with the imposition of these added conditions, the matter is still understated since the problem falls in the category of an ill-posed problem. The goal here will be somewhat more modest.

The following development will assume sufficient information on the coefficient c to render the sinc methodology (Theorem 3.5) of Chapter 3 applicable and leave to future development the more general setting. In particular, assume that the function c is in $B(D_E)$ and that c vanishes at 0 and 1 so that the interpolatory solution

$$c_{m_c}(x) = \sum_{j=-M_c}^{N_c} c_j S(j,h) \circ \phi(x), \quad m_c = M_c + N_c + 1 \tag{5.70}$$

provides an accurate approximation of c.

Orthogonalizing the residual with respect to the basis functions $S(k,h) \circ \phi(x)$ and omitting the error resulting from the application of the approximate inner products in Theorem 4.4 yields

$$\begin{aligned}
0 &= \int_0^1 \left[f(x) + \frac{d}{dx}\left(c_{m_c}(x)\frac{du}{dx}\right)\right] S(k,h) \circ \phi(x)\, w(x)\,dx \\
&= h\left(\frac{fw}{\phi'}\right)(x_k) + \left\{\left(\frac{c_{m_c}u'[S(k,h)\circ\phi]\,w}{\phi'}\right)(x)\Big|_0^1\right. \\
&\quad \left. - \int_0^1 c_{m_c}(x)u'(x)\left(S(k,h)\circ\phi(x)w(x)\right)'\,dx\right\}
\end{aligned}$$

which upon rearrangement reads

$$\int_0^1 c_{m_{\dot{c}}}(x)u'(x)\left([S(k,h)\circ\phi]w\right)'(x)dx = h\left(\frac{fw}{\phi'}\right)(x_k)$$

if it is assumed that the integrated term vanishes,

$$\left(\frac{c_{m_c}u'[S(k,h)\circ\phi]w}{\phi'}\right)(x)\Big|_0^1 = 0.$$

Using (5.70), followed by an integration by parts, leads to

$$-\sum_{j=-M_c}^{N_c} c_j \int_0^1 u(x)\left\{S(j,h)\circ\phi(x)\left(S(k,h)\circ\phi(x)\,w(x)\right)'\right\}' dx$$

$$= h\left(\frac{fw}{\phi'}\right)(x_k). \tag{5.71}$$

Here the term

$$u(x)\left\{S(j,h)\circ\phi(x)\left(S(k,h)\circ\phi(x)w(x)\right)'\right\}\Big|_0^1$$

vanishes due to the homogeneous boundary conditions in (5.69).

Assume that the solution u of (5.69) is in $B(D_E)$ and replace u in (5.71) by

$$u_m(x) = \sum_{\ell=-M}^{N} u_\ell S(\ell,h)\circ\phi(x), \quad m = M+N+1.$$

Using the point evaluation quadrature in (4.30) to approximate the resulting integrals leads to the approximate identity

$$\int_0^1 S(\ell,h)\circ\phi(x)\left\{S(j,h)\circ\phi(x)\left[S(k,h)\circ\phi(x)w(x)\right]'\right\}' dx$$

$$= h\left(\frac{\left\{[S(j,h)\circ\phi]\left([S(k,h)\circ\phi]\,w\right)'\right\}'}{\phi'}\right)(x_\ell).$$

Substituting these approximate integrals in the left-hand side of (5.71) and interchanging the two finite sums yields

$$-h \sum_{\ell=-M}^{N} \left(\frac{([S(k,h) \circ \phi] \, w)''}{\phi'} \right) (x_\ell) \sum_{j=-M_c}^{N_c} \delta_{j\ell}^{(0)} u_\ell c_j$$

$$-h \sum_{\ell=-M}^{N} \left(\frac{([S(k,h) \circ \phi] \, w)'}{\phi'} \right) (x_\ell) \sum_{j=-M_c}^{N_c} \delta_{j\ell}^{(1)} \phi'(x_\ell) u_\ell c_j \quad (5.72)$$

$$= h \left(\frac{fw}{\phi'} \right) (x_k).$$

Taking $m_c = m$ in (5.72) and letting k vary between $-M$ and N in (5.72) leads to the matrix system

$$\{ \mathcal{A}(w) D(\phi' w) D(u_m) + \mathcal{B}(w) D(w) D(u_m) F \} \vec{c} \quad (5.73)$$

$$= D \left(\frac{wf}{\phi'} \right) \vec{1}$$

where the matrices $\mathcal{A}(w)$ and $\mathcal{B}(w)$ are defined in (4.40) and (4.79), respectively. The matrix F is given by

$$F = -\frac{1}{h} D(\phi') I^{(1)}.$$

Depending on the properties of the true solution $u(x)$ in (5.69), there are various options for the selection of the weight function $w(x)$. The selection $w(x) = (\phi'(x))^{-1/2}$ in (5.72) will be used due to the simplicity of the coefficient matrix. Since $(\sqrt{\phi'})' + \phi' (1/\sqrt{\phi'})' \equiv 0$, the equation in (5.73) takes the form

$$\left\{ \left[\mathcal{A} \left(\frac{1}{\sqrt{\phi'}} \right) \right] D \left((\phi')^{1/2} u_m \right) + \left[\mathcal{B} \left(\frac{1}{\sqrt{\phi'}} \right) \right] D \left((\phi')^{-1/2} u_m \right) F \right\} \vec{c} \quad (5.74)$$

$$= D \left(\frac{f}{(\phi')^{3/2}} \right) \vec{1}.$$

The performance of this matrix system to find c is illustrated in Example 5.5.

Notice that one could also isolate the coefficients for the approximate solution u_m in (5.73). If the notation $\mathcal{D}(\vec{v})$ designates the diagonal matrix whose kk-th entry is the k-th component of the vector $\vec{v}$, then the identities

$$D(u_m)\vec{c} = D(c_{m_c})\vec{u}$$

and

$$D(u_m)F\vec{c} = D(F\vec{c})\vec{u}$$

allow the solution of (5.73) for the coefficients u_ℓ in the system

$$\left\{ \mathcal{A}\left(\frac{1}{\sqrt{\phi'}}\right) D\left((\phi')^{1/2} c_{m_c}\right) + \mathcal{B}\left(\frac{1}{\sqrt{\phi'}}\right) D\left((\phi')^{-1/2}\right) \mathcal{D}(F\vec{c}) \right\} \vec{u} \tag{5.75}$$

$$= D\left(\frac{f}{(\phi')^{3/2}}\right) \vec{1}.$$

This is the forward form (solve for u_m) of (5.74). This system is different from the system recorded in (5.68), which was obtained formally from the work in Section 5.1. The system listed in (5.75) is the system used in the recovery of the coefficient in time-dependent parabolic partial differential equations [12] and [15]. Further work along these lines involving the recovery of material parameters in Euler–Bernoulli beam models (fourth-order partial differential equations) can be found in [20], [21], and [22]. In the present steady-state case the performance is illustrated in the following example.

Example 5.5 Consider the differential equation

$$-\frac{d}{dx}\left(c(x)\frac{du}{dx}\right) = -\pi \cos(\pi x)(1 - 2x) - 2\sin(\pi x), \quad 0 < x < 1 \tag{5.76}$$

$$u(0) = u(1) = 0.$$

The true solution of this problem with the coefficient $c(x) = \sin(\pi x)$ is given by $u(x) = x(1 - x)$. Letting $\alpha = \beta = 1$ gives $M_c = N_c$ and thus $h = \pi/\sqrt{2M_c}$ is used. The errors in the coefficient $c(x)$ are given in Table 5.4 below, reported as $\|E_S(h)\|$.

STEADY PROBLEMS

$M_c = N_c$	$h = \pi/\sqrt{2M_c}$	$\|E_S(h)\|$
4	1.111	$.145 \times 10^{-1}$
8	0.785	$.109 \times 10^{-2}$
16	0.555	$.339 \times 10^{-2}$
32	0.393	$.173 \times 10^{-5}$

Table 5.4 Error on the sinc grid between $c(x)$ and $c_{m_c}(x)$ for (5.76).

■

Before turning to a speculative example, it is convenient to record here the discrete method of the present section for the problem

$$L(c)u(x) + q(x)u(x) = f(x).$$

The only alteration to the system (5.75) (for an arbitrary weight $w(x)$) is the addition of the diagonal matrix $D(wq/(\phi'))$, that is,

$$\left\{ \mathcal{A}(w)D(\phi' w c_{m_c}) + \mathcal{B}(w)D(w)\mathcal{D}(F\vec{c}) + D\left(\frac{wq}{\phi'}\right) \right\} \vec{u} \quad (5.77)$$

$$= D\left(\frac{wf}{\phi'}\right) \vec{1}.$$

As a closing muse, consider using the system (5.74) to solve for $c(x)$ in the problem

$$-\frac{d}{dx}\left(c(x)\frac{du}{dx}\right) = 1 + 4x, \quad 0 < x < 1$$

$$u(0) = u(1) = 0 \quad (5.78)$$

where $u(x) = x(1-x)$. Notice that $c(x)$, which is given for any δ by

$$c_\delta(x) = 1 + x + \frac{\delta}{1 - 2x},$$

does not vanish at $x = 1$. The forward solver works very well in the computation of $u(x)$, but the system (5.74) is not cognizant of the behavior of $c(x)$ at 0 or 1. The discretization is forced to pick out some solution from this family of solutions. If one solves the problem (5.69) for the coefficient $c(x)$ by applying the method in (5.77) to determine the coefficients c_j in (5.70), the results show that the error between $c_0(x)$ and the sinc approximation to c_{m_c} is concentrated around $x = 1/2$ (Exercise 5.6). It appears that the method is attempting to select the nonsingular solution ($\delta = 0$) in the family of available solutions but is confused in the vicinity of $x = 1/2$. The problem is, as stated, underdetermined.

5.4 The Poisson Problem

The Sinc-Galerkin method applied to the Poisson problem in two and three variables is discussed in this section. The basis, trial function, and inner product are all extensions of those used in Section 4.2 for two-point boundary value problems. Hence the major obligation of this section is to catalogue the appropriate discrete systems needed to solve for the approximate solution. The discussion will thus focus on equivalent formulations of the discrete system; in the first the unknowns are treated as a matrix, whereas in the second they are treated as a vector. Though equivalent, each formulation has its computational advantages and disadvantages. By referring to the material in Appendix A.3 the various formulations and their distinctive features will be analyzed. The somewhat lengthy presentation is due to the decision to include the details of both the two- and three-dimensional problems. In the former case, there is included a manner of arriving at the discrete system for the two-dimensional problem that has proved instructive but does not seem to have been discussed too much elsewhere. Alternative sinc method approaches for the integral equation formulation are described in [23].

Consider the equilibrium problem

$$\begin{aligned}
-\Delta^{(2)} u(x,y) &\equiv -(u_{xx}(x,y) + u_{yy}(x,y)) = f(x,y), \\
&(x,y) \in S = (0,1) \times (0,1) \quad (5.79) \\
u(x,y) &= 0, \quad (x,y) \in \partial S.
\end{aligned}$$

STEADY PROBLEMS

The assumed approximate solution takes the form

$$u_{m_x,m_y}(x,y) = \sum_{j=-M_y}^{N_y} \sum_{i=-M_x}^{N_x} u_{ij} S_{ij}(x,y) \qquad (5.80)$$

where

$$m_x = M_x + N_x + 1,$$
$$m_y = M_y + N_y + 1.$$

The basis functions $\{S_{ij}(x,y)\}$ for $-M_x \leq i \leq N_x$, $-M_y \leq j \leq N_y$ are given as simple product basis functions of the one-dimensional sinc basis

$$S_{ij}(x,y) = [S(i,h_x) \circ \phi(x)][S(j,h_y) \circ \phi(y)]$$

where the conformal map

$$\phi(z) = \ell n \left(\frac{z}{1-z}\right)$$

will be used exclusively in this section. The slight abuse in notation will cause no confusion as the independent variable x (or y) will be specified by subscripting where appropriate. For example, these maps will be distinguished by ϕ_x (or ϕ_y) in the discrete system (5.83), where the matrix dimensions need to be distinguished. A representative member of the mapped sinc basis is shown in Figure 5.1.

The inner product is defined by

$$(f,g) = \int_0^1 \int_0^1 f(x,y) g(x,y) v(x) w(y) dx dy$$

where the product $v(x)w(y)$ plays the role of a weight function. Following the development in Chapter 4, the discrete system for the coefficients $\{u_{ij}\}$ in (5.80) is obtained by orthogonalizing the residual with respect to the inner product

$$\left(-\Delta^{(2)} u - f, S_{k\ell}\right) = 0 \qquad (5.81)$$

where $-M_x \leq k \leq N_x$ and $-M_y \leq \ell \leq N_y$.

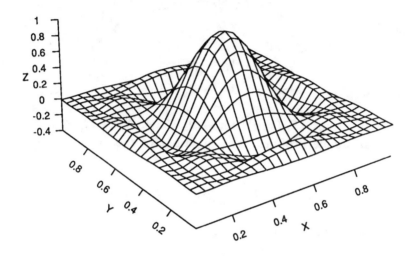

Figure 5.1 The sinc basis on $S = (0,1) \times (0,1)$ for $i = j = 0$ and $h = \pi/4$.

In a manner analogous to that of Section 4.2, the derivatives are removed from u by applying Green's identity. Specifically,

$$\int_0^1 \int_0^1 \Delta^{(2)} u(x,y) S(k,h_x) \circ \phi(x) S(\ell, h_y) \circ \phi(y) v(x) w(y) dx dy$$

$$= \int_0^1 \int_0^1 u(x,y) \Delta^{(2)} [S(k,h_x) \circ \phi(x) S(\ell, h_y) \circ \phi(y) v(x) w(y)] dx dy$$

$$- B_{T_2}$$

where

$$B_{T_2} = \int_{\partial S} \left\{ u(x,y) \frac{\partial}{\partial n} [S(k,h_x) \circ \phi(x) v(x) S(\ell, h_y) \circ \phi(y) w(y)] \right.$$

$$\left. - S(k,h_x) \circ \phi(x) v(x) S(\ell, h_y) \circ \phi(y) w(y) \frac{\partial u}{\partial n}(x,y) \right\} d\sigma$$

STEADY PROBLEMS 197

and n is the outward normal direction to S. Assuming $B_{T_2} = 0$, (5.81) yields

$$\int_0^1 \int_0^1 -u(x,y)S(\ell,h_y) \circ \phi(y)w(y)[S(k,h_x) \circ \phi(x)v(x)]'' dxdy$$
$$+ \int_0^1 \int_0^1 -u(x,y)S(k,h_x) \circ \phi(x)v(x)[S(\ell,h_y) \circ \phi(y)w(y)]'' dxdy$$
$$= \int_0^1 \int_0^1 f(x,y)S(k,h_x) \circ \phi(x)v(x)S(\ell,h_y) \circ \phi(y)w(y) dxdy.$$

Applying the quadrature rule (Theorem 3.8) to the iterated integrals, deleting the error terms, replacing $u(x_p, y_q)$ by u_{pq}, and dividing by $h_x h_y$ yields the discrete sinc system

$$\frac{w(y_\ell)}{\phi'(y_\ell)} \sum_{p=-M_x}^{N_x} \left[-\frac{1}{h_x^2} \delta_{kp}^{(2)} \phi'(x_p) v(x_p) \right.$$
$$- \frac{1}{h_x} \delta_{kp}^{(1)} \left(\frac{\phi''(x_p) v(x_p)}{\phi'(x_p)} + 2v'(x_p) \right) \quad (5.82)$$
$$\left. - \delta_{kp}^{(0)} \frac{v''(x_p)}{\phi'(x_p)} \right] u_{p\ell} + \frac{v(x_k)}{\phi'(x_k)} \sum_{q=-M_y}^{N_y} \left[-\frac{1}{h_y^2} \delta_{\ell q}^{(2)} \phi'(y_q) w(y_q) \right.$$
$$- \frac{1}{h_y} \delta_{\ell q}^{(1)} \left(\frac{\phi''(y_q) w(y_q)}{\phi'(y_q)} + 2w'(y_q) \right) - \delta_{\ell q}^{(0)} \frac{w''(y_q)}{\phi'(y_q)} \bigg] u_{kq}$$
$$= \frac{f(x_k, y_\ell) v(x_k) w(y_\ell)}{\phi'(x_k) \phi'(y_\ell)}.$$

This system is identical to the system generated by orthogonalizing the residual via $(-\Delta^{(2)} u_{m_x,m_y} - f, S_{k\ell}) = 0$. Recall the notation $I_{m_x}^{(\ell)}$, $\ell = 0, 1, 2$, where the $m_x \times m_x$ matrices $I^{(\ell)}$, with jk-th entry $\delta_{jk}^{(\ell)}$ are given in (4.14)–(4.16). Further, $D(g_x)$ is an $m_x \times m_x$ diagonal matrix whose diagonal entries are $g(x_{-M_x}), \ldots, g(x_{N_x})$. The matrices $I_{m_y}^{(\ell)}$, $\ell = 0, 1, 2$ and $D(g_y)$ are similarly defined though of size $m_y \times m_y$. Lastly, the $m_x \times m_y$ matrices $U^{(2)}$ and $F^{(2)}$ have jk-th entries given by u_{jk} and $f(x_j, y_k) = f(e^{jh_x}/(e^{jh_x}+1), e^{kh_y}/(e^{kh_y}+1))$, respectively. Introducing this notation in (5.82) leads to the matrix form

$$\left[-\frac{1}{h_x^2} I_{m_x}^{(2)} D(\phi_x' v) - \frac{1}{h_x} I_{m_x}^{(1)} D\left(\frac{\phi_x'' v}{\phi_x'} + 2v'\right)\right.$$

$$\left. - I_{m_x}^{(0)} D\left(\frac{v''}{\phi_x'}\right)\right] U^{(2)} D\left(\frac{w}{\phi_y'}\right) + D\left(\frac{v}{\phi_x'}\right) U^{(2)} \left[-\frac{1}{h_y^2} I_{m_y}^{(2)} D(\phi_y' w)\right.$$

$$\left. -\frac{1}{h_y} I_{m_y}^{(1)} D\left(\frac{\phi_y'' w}{\phi_y'} + 2w'\right) - I_{m_y}^{(0)} D\left(\frac{w''}{\phi_y'}\right)\right]^T$$

$$= D\left(\frac{v}{\phi_x'}\right) F^{(2)} D\left(\frac{w}{\phi_y'}\right).$$

Premultiplying by $D(\phi_x')$ and postmultiplying by $D(\phi_y')$ yields the equivalent system

$$D(\phi_x')\left[-\frac{1}{h_x^2} I^{(2)} D(\phi_x' v) - \frac{1}{h_x} I^{(1)} D\left(\frac{\phi_x'' v}{\phi_x'} + 2v'\right)\right.$$

$$\left. - D\left(\frac{v''}{\phi_x'}\right)\right] U^{(2)} D(w) + D(v) U^{(2)} \left[-\frac{1}{h_y^2} I^{(2)} D(\phi_y' w)\right.$$

$$\left. -\frac{1}{h_y} I^{(1)} D\left(\frac{\phi_y'' w}{\phi_y'} + 2w'\right) - D\left(\frac{w''}{\phi_y'}\right)\right]^T D(\phi_y') \quad (5.83)$$

$$= D(v) F^{(2)} D(w).$$

The subscripts m_x and m_y have been dropped and the matrix sizes are implied by the dimensions of the diagonal matrices $D(\phi_q)$, $q = x$ or y. The representation of the system is simplified upon recalling the definition of the matrix $\mathcal{A}(w)$ in (4.40). In this notation the system in (5.83) is the Sylvester equation (A.31)

$$D(\phi_x')\mathcal{A}(v)D(\phi_x')V^{(2)} + V^{(2)}[D(\phi_y')\mathcal{A}(w)D(\phi_y')]^T = G^{(2)} \quad (5.84)$$

where the weighted coefficients are defined by

$$V^{(2)} = D(v)U^{(2)}D(w) \quad (5.85)$$

and the right-hand side is given by

$$G^{(2)} = D(v)F^{(2)}D(w). \quad (5.86)$$

Using the weight function
$$v(x)w(y) = \frac{1}{\phi'(x)\phi'(y)}, \tag{5.87}$$
many examples were numerically tested in [3]. With this selection the equation (5.84) becomes
$$D(\phi'_x)\mathcal{A}((\phi'_x)^{-1})D(\phi'_x)V^{(2)} + V^{(2)}[D(\phi'_y)\mathcal{A}((\phi'_y)^{-1})D(\phi'_y)]^T = G^{(2)}.$$

Here $V^{(2)}$ and $G^{(2)}$ are given in (5.85) and (5.86), respectively, with $v = 1/\phi'_x$ and $w = 1/\phi'_y$. The components of this discrete system are given in Table 4.1. Note that, as in the case of the one-dimensional problem, the matrices $\mathcal{A}(1/(\phi'_q))$ ($q = x$ or y) are not symmetric.

If instead of the weight function $v(x)w(y) = (\phi'(x)\phi'(y))^{-1}$ one selects [13]
$$v(x)w(y) = \frac{1}{\sqrt{\phi'(x)\phi'(y)}} \tag{5.88}$$
then (5.84) becomes
$$D(\phi'_x)\mathcal{A}((\phi'_x)^{-1/2})D(\phi'_x)V^{(2)} + V^{(2)}[D(\phi'_y)\mathcal{A}((\phi'_y)^{-1/2})D(\phi'_y)]^T$$
$$= G^{(2)}.$$

Upon defining
$$\begin{aligned}A_x &\equiv D(\phi'_x)\mathcal{A}((\phi'_x)^{-1/2})D(\phi'_x) \\ &= D(\phi'_x)\left[-\frac{1}{h_x^2}I^{(2)} + D\left(\frac{-1}{(\phi'_x)^{3/2}}\left(\frac{1}{\sqrt{\phi'_x}}\right)''\right)\right]D(\phi'_x) \\ &= D(\phi'_x)\left[-\frac{1}{h_x^2}I^{(2)} + D\left(\frac{1}{4}\right)\right]D(\phi'_x)\end{aligned} \tag{5.89}$$

with an analogous definition for A_y the system in (5.84), with weight function given in (5.88), may be written in the form
$$A_x V^{(2)} + V^{(2)} A_y^T = G^{(2)} \tag{5.90}$$
where from (5.85) and (5.86)
$$V^{(2)} = D((\phi'_x)^{-1/2})U^{(2)}D((\phi'_y)^{-1/2})$$

and
$$G^{(2)} = D((\phi'_x)^{-1/2})F^{(2)}D((\phi'_y)^{-1/2}),$$
respectively. Here the matrices A_x and A_y are symmetric, so the transpose in (5.90) is unnecessary. The second equality in (5.89) is the reason the problem was posed on $(0,1)$. If the original problem had been posed on the interval $(a_1, b_1) \times (a_2, b_2)$ then one would need only alter the definition of the nodes and map $\phi(x)$ from Table 4.7, not the structure of the matrices in the system.

An alternative way to arrive at (5.84) is obtained by considering the two one-dimensional problems

$$-u_{xx}(x, y_\ell) = f(x, y_\ell) + u_{yy}(x, y_\ell) \equiv f_\ell(x), \quad 0 < x < 1 \tag{5.91}$$

$$u(0, y_\ell) = u(1, y_\ell) = 0, \quad -M_y \le \ell \le N_y$$

and

$$-u_{yy}(x_k, y) = f(x_k, y) + u_{xx}(x_k, y) \equiv g_k(y), \quad 0 < y < 1 \tag{5.92}$$

$$u(x_k, 0) = u(x_k, 1) = 0, \quad -M_x \le k \le N_x.$$

Define the one-dimensional approximations

$$u_{m_x, m_y}(x, y_\ell) = \sum_{p=-M_x}^{N_x} u_{p\ell} S(p, h_x) \circ \phi(x), \quad -M_y \le \ell \le N_y$$

and

$$u_{m_x, m_y}(x_k, y) = \sum_{q=-M_y}^{N_y} u_{kq} S(q, h_y) \circ \phi(y), \quad -M_x \le k \le N_x$$

to the solutions of (5.91) and (5.92), respectively. The situation now is completely analogous to that in Section 4.2 for the one-dimensional problem (4.1) with $p(x) \equiv q(x) \equiv 0$. The discrete systems for $\{u_{p\ell}\}_{p=-M_x}^{N_x}$ $(-M_y \le \ell \le N_y)$ and $\{u_{kq}\}_{q=-M_y}^{N_y}$ $(-M_x \le k \le N_x)$ are determined from (4.39), where the weight functions used are $v(x)$ or $w(y)$, respectively. Thus for $-M_y \le \ell \le N_y$

$$\mathcal{A}(v) D(\phi'_x v) \vec{u}^{(\ell)} = D\left(\frac{f_\ell v}{\phi'_x}\right) \vec{1}$$

STEADY PROBLEMS

and for $-M_x \leq k \leq N_x$

$$\mathcal{A}(w)D(\phi'_y w)\vec{u}_{(k)} = D\left(\frac{g_k w}{\phi'_y}\right)\vec{1}.$$

Here $\vec{u}^{(\ell)} = (u_{-M_x,\ell}, \ldots, u_{N_x,\ell})^T$ and $\vec{u}_{(k)} = (u_{k,-M_y}, \ldots, u_{k,N_y})^T$. Stacking each of these problems up as k and ℓ vary and premultiplying by $D(\phi'_x/v)$ and $D(\phi'_y/w)$, respectively, produces the two systems

$$D\left(\frac{\phi'_x}{v}\right)\mathcal{A}(v)D(\phi'_x v)U = F_1$$

and

$$D\left(\frac{\phi'_y}{w}\right)\mathcal{A}(w)D(\phi'_y w)V = F_2$$

where

$$U = \begin{bmatrix} u_{-M_x,-M_y} & \cdots & u_{-M_x,N_y} \\ \vdots & & \vdots \\ u_{N_x,-M_y} & \cdots & u_{N_x,N_y} \end{bmatrix}_{m_x \times m_y},$$

and $V = U^T$. The right-hand sides are given by

$$F_1 = \begin{bmatrix} (f+u_{yy})(x_{-M_x}, y_{-M_y}) & \cdots & (f+u_{yy})(x_{-M_x}, y_{N_y}) \\ \vdots & & \vdots \\ (f+u_{yy})(x_{N_x}, y_{-M_y}) & \cdots & (f+u_{yy})(x_{N_x}, y_{N_y}) \end{bmatrix}$$
$$\equiv F + U_{yy}$$

and

$$F_2 = \begin{bmatrix} (f+u_{xx})(x_{-M_x}, y_{-M_y}) & \cdots & (f+u_{xx})(x_{N_x}, y_{-M_y}) \\ \vdots & & \vdots \\ (f+u_{xx})(x_{-M_x}, y_{N_y}) & \cdots & (f+u_{xx})(x_{N_x}, y_{N_y}) \end{bmatrix}$$
$$\equiv F^T + U_{xx}^T.$$

Transposing the equation for V and adding to this transposition the U equation yields

$$D\left(\frac{\phi'_x}{v}\right)\mathcal{A}(v)D(\phi'_x v)U + U\left\{D\left(\frac{\phi'_y}{w}\right)\mathcal{A}(w)D(\phi'_y w)\right\}^T$$
$$= F_1 + F_2^T$$
$$= F + U_{yy} + F + U_{xx}$$
$$= F$$

since $-U_{xx} - U_{yy} = F$. This is the same system as (5.84), where $U = U^{(2)}$ and $F = F^{(2)}$. Thus one may consider the two one-dimensional problems instead of one two-dimensional problem to arrive at the discrete system (5.84). Furthermore, this gives insight, via analogy with the one-dimensional problem, into the conditions that yield the parameter selections.

For the weight function specified in (5.87) application of Theorem 4.7 to each of the problems (5.91) and (5.92) requires that for $q = x$ or y, $f/\phi'_q \in B(D_E)$ and $uF_q \in B(D_E)$, where

$$F_q = (1/\phi'_q)'', \quad (\phi''_q/\phi'_q), \quad \phi'_q.$$

Further assume that there are positive constants $C(y), C(x), \alpha, \beta, \zeta$, and η so that

$$|u(x,y)| \leq C(y) \begin{cases} x^\alpha, & x \in \left(0, \frac{1}{2}\right) \\ (1-x)^\beta, & x \in \left[\frac{1}{2}, 1\right) \end{cases} \quad (5.93)$$

and

$$|u(x,y)| \leq C(x) \begin{cases} y^\zeta, & y \in \left(0, \frac{1}{2}\right) \\ (1-y)^\eta, & y \in \left[\frac{1}{2}, 1\right) \end{cases}. \quad (5.94)$$

Then choosing

$$N_x = \left[\left|\frac{\alpha}{\beta} M_x + 1\right|\right], \quad h_x = \left(\frac{\pi d}{\alpha M_x}\right)^{1/2} \quad (5.95)$$

and

$$N_y = \left[\left|\frac{\zeta}{\eta} M_y + 1\right|\right], \quad h_y = \left(\frac{\pi d}{\zeta M_y}\right)^{1/2}$$

results in

$$\|u - u_{m_x, m_y}\|_\infty \leq \hat{C}(y) M_x^2 \exp\left(-(\pi d \alpha M_x)^{1/2}\right)$$

and

$$\|u - u_{m_x, m_y}\|_\infty \leq \hat{C}(x) M_y^2 \exp\left(-(\pi d \zeta M_y)^{1/2}\right).$$

Now choose

$$h \equiv h_x = h_y \quad (5.96)$$

which yields

$$M_y = \left[\left|\frac{\alpha}{\zeta} M_x + 1\right|\right], \quad N_y = \left[\left|\frac{\alpha}{\eta} M_x + 1\right|\right]. \quad (5.97)$$

Then combining (5.93) and (5.94) as

$$|u(x,y)| \leq C x^\alpha (1-x)^\beta y^\zeta (1-y)^\eta, \quad (x,y) \in S \quad (5.98)$$

and making the selections (5.95), (5.96), and (5.97) leads to

$$\|u - u_{m_x,m_y}\|_\infty \leq K M_x^2 \exp\left(-(\pi d\alpha M_x)^{1/2}\right). \quad (5.99)$$

For the weight function $v(x)w(y) = (\phi'(x)\phi'(y))^{-1/2}$ in (5.88) the application of Theorem 4.11 to each of the problems (5.91) and (5.92) requires that ($q = x$ or y) $f/\sqrt{\phi_q'}$ and $uF_q \in B(D_E)$, where

$$F_q = \sqrt{\phi_q'}\left(1/\phi_q'\right)'', \quad (\phi_q')^{3/2}, \quad \phi_q''/\sqrt{\phi_q'}, \quad \left(1/\sqrt{\phi_q'}\right)''.$$

Furthermore, assume that for $(x,y) \in S$

$$|u(x,y)| \leq C x^{\alpha_s+1/2}(1-x)^{\beta_s+1/2} y^{\zeta_s+1/2}(1-y)^{\eta_s+1/2}.$$

Then making the selections (5.95), (5.96), and (5.97) (using α_s, β_s, ζ_s, η_s, and h_s) leads to

$$\|u - u^s_{m_x,m_y}\|_\infty \leq K_s M_x^2 \exp\left(-(\pi d\alpha_s M_x)^{1/2}\right)$$

where

$$u^s_{m_x,m_y}(x,y) = \sum_{j=-M_y}^{N_y} \sum_{i=-M_x}^{N_x} u^s_{ij} S_{ij}(x,y). \quad (5.100)$$

The formulation (5.84) may be rewritten as the large sparse discrete system using the Kronecker sum notation and the concatenation (co) of matrices (Definitions A.26 and A.29)

$$\mathcal{A}^{(2)} \mathrm{co}(V^{(2)}) = \mathrm{co}(G^{(2)}) \quad (5.101)$$

where the $m_x m_y \times m_x m_y$ matrix $\mathcal{A}^{(2)}$ is

$$\mathcal{A}^{(2)} \equiv I_{m_y} \otimes D(\phi'_x)\mathcal{A}(v)D(\phi'_x) \\ + D(\phi'_y)\mathcal{A}(w)D(\phi'_y) \otimes I_{m_x}. \quad (5.102)$$

The vectors

$$\begin{aligned} \operatorname{co}(V^{(2)}) &= \operatorname{co}\left(D(v)U^{(2)}D(w)\right) \\ &= [D(w) \otimes D(v)]\operatorname{co}(U^{(2)}) \\ &\equiv D_{wv}\operatorname{co}(U^{(2)}) \end{aligned}$$

and

$$\operatorname{co}(G^{(2)}) = D_{wv}\operatorname{co}(F^{(2)})$$

have dimension $m_x m_y \times 1$, where Theorem A.22 has been used. Note that $\operatorname{co}(U^{(2)})$ corresponds to a natural or lexicographic ordering of the sinc gridpoints from left to right, bottom to top. Thus (x_i, y_j) follows (x_k, y_ℓ) if $y_j > y_\ell$ or if $y_j = y_\ell$ and $x_i > x_k$.

For purposes of illustrating a solution method for the general equation (5.84) or equivalently (5.101), assume that $D(\phi'_x)\mathcal{A}(v)D(\phi'_x)$ and $[D(\phi'_y)\mathcal{A}(w)D(\phi'_y)]^T$ are diagonalizable (this will depend on the choice of weight function). Diagonalizability guarantees two nonsingular matrices P and Q such that

$$P^{-1}D(\phi'_x)\mathcal{A}(v)D(\phi'_x)P = \Lambda_x$$

and

$$Q^{-1}[D(\phi'_y)\mathcal{A}(w)D(\phi'_y)]^T Q = \Lambda_y.$$

From Appendix A.3, (5.84) is equivalent to

$$\Lambda_x W^{(2)} + W^{(2)}\Lambda_y = H^{(2)} \tag{5.103}$$

where

$$W^{(2)} = P^{-1}V^{(2)}Q \tag{5.104}$$

and

$$H^{(2)} = P^{-1}G^{(2)}Q.$$

Thus if the spectrums of the matrices are denoted by

$$\sigma\left(D(\phi'_x)\mathcal{A}(v)D(\phi'_x)\right) = \{(\lambda_x)_i\}_{i=-M_x}^{N_x}$$

and

$$\sigma\left(D(\phi'_y)\mathcal{A}(w)D(\phi'_y)\right) = \{(\lambda_y)_j\}_{j=-M_y}^{N_y}$$

STEADY PROBLEMS

then (5.103) has the component solution

$$w_{ij} = \frac{h_{ij}}{(\lambda_x)_i + (\lambda_y)_j}, \quad -M_x \leq i \leq N_x, \quad -M_y \leq j \leq N_y. \quad (5.105)$$

Using (5.104) and (5.85), $U^{(2)}$ is recovered from $W^{(2)}$ by

$$U^{(2)} = D\left(\frac{1}{v}\right) P W^{(2)} Q^{-1} D\left(\frac{1}{w}\right).$$

The equivalent treatment of (5.101) is to multiply on the left by $Q^T \otimes P^{-1}$ and note that

$$\begin{aligned}(Q^T \otimes P^{-1})\mathrm{co}(V^{(2)}) &= \mathrm{co}(P^{-1} V^{(2)} Q) \\ &= \mathrm{co}(W^{(2)})\end{aligned}$$

and

$$\begin{aligned}(Q^T \otimes P^{-1})\mathrm{co}(G^{(2)}) &= \mathrm{co}(P^{-1} G^{(2)} Q) \\ &= \mathrm{co}(H^{(2)}).\end{aligned}$$

This yields

$$[I_{m_y} \otimes \Lambda_x + \Lambda_y \otimes I_{m_x}]\mathrm{co}(W^{(2)}) = \mathrm{co}(H^{(2)}) \quad (5.106)$$

and the solution of (5.106) is given by (5.105).

If the weight function in (5.88) is used, then the coefficient matrix in (5.89) is symmetric and matters are a bit simpler. In this case the equation (5.90) reads

$$A_x V^{(2)} + V^{(2)} A_y = G^{(2)}$$

and the matrices P and Q above may be taken to be orthogonal. That is, the above similarities are replaced by

$$P^T A_x P = \Lambda_x \quad (5.107)$$

and

$$Q^T A_y Q = \Lambda_y. \quad (5.108)$$

This provides a simplification as no inverses need be directly calculated ($P^{-1} = P^T$, $Q^{-1} = Q^T$). Moreover, if $M_x = N_x$ and $M_y = N_y$, then the diagonalizers P and Q are the same (Exercise 5.7).

The coefficient matrix $\mathcal{A}^{(2)}$ has a readily discerned block structure with blocks $\mathcal{A}^{(2)}_{ij}$ given by

$$\mathcal{A}^{(2)}_{ij} = \delta^{(0)}_{ij} D(\phi'_x) A(v) D(\phi'_x) + \left(D(\phi'_y) A(w) D(\phi'_y)\right)_{ij} I_{m_x}$$

with $(\cdot)_{ij}$ denoting the ij-th element of the given matrix. This block structure is graphically illustrated in Figure 5.2.

$$\begin{bmatrix} X\,X\,X & X & & X & & X & & \\ X\,X\,X & & X & & X & & X & \\ X\,X\,X & & & X & & X & & X \\ X & X\,X\,X & & X & & X & & \\ & X & X\,X\,X & & X & & X & \\ & & X\,X\,X\,X & & & X & & X \\ X & & X & X\,X\,X & X & & & \\ & X & & X & X\,X\,X & & X & \\ & & X & & X\,X\,X\,X & & & X \\ X & & X & & X & X\,X\,X & & \\ & X & & X & & X & X\,X\,X & \\ & & X & & X & & X\,X\,X\,X \end{bmatrix}$$

Figure 5.2 Nonzero elements in the matrix $\mathcal{A}^{(2)}$ for $m_x = 3$, $m_y = 4$.

The matrix $\mathcal{A}^{(2)}$ is very sparse and regularly structured. This structure is accessibly revealed by an inspection of the Kronecker sum

$$\mathcal{A}^{(2)} = I_{m_y} \otimes A_x + A_y \otimes I_{m_x}. \tag{5.109}$$

Due to the structural equivalence of the systems (only the weight function has been changed), Figure 5.2 also represents the more general matrix in (5.102).

The method of solution outlined above, although illustrative, has analytic and computational disadvantages. Analytically, it requires the diagonalizability of the matrices $D(\phi'_x)A(v)D(\phi'_x)$ and $D(\phi'_y)A(w)D(\phi'_y)$. For the case of (5.90) there is no problem due to symmetry. In the case of (5.84) the situation parallels the discussion following Theorem 4.19 for the one-dimensional problem. For h sufficiently small the matrix $I^{(2)}$, a diagonalizable matrix, is the dominant matrix in $\mathcal{A}(1/\phi')$; whether the sufficiently small h is in communion with the mesh selection in (5.95) is not known. The many examples tested in [3], which were computed by this procedure, would argue favorably that there is no problem with the assumption. Even if the diagonalizability assumption is valid, a procedure based on a form of block Gauss–Jordan elimination is suggested by the sparse structure in Figure 5.2.

Perhaps more in the vein of the diagonalizing procedure is a method based on Schur's Theorem A.6 to obtain unitary Q and R with the properties

$$Q^*D(\phi'_x)\mathcal{A}(v)D(\phi'_x)Q = S \tag{5.110}$$

and

$$R^*D(\phi'_y)\mathcal{A}(w)D(\phi'_y)R = T \tag{5.111}$$

where S and T are upper triangular. Then (5.84) is equivalent to

$$SW^{(2)} + W^{(2)}T^* = H^{(2)} \tag{5.112}$$

where

$$W^{(2)} = Q^*V^{(2)}R$$

and

$$H^{(2)} = Q^*G^{(2)}R.$$

Note that T^* is lower triangular, so viewing (5.112) column by column yields (remember $A\vec{x} = \sum_{j=1}^n x_j A_j$, where A_j are the columns of A)

$$S\vec{w}_k + \sum_{j=k}^{m_y} t^*_{jk}\vec{w}_j = \vec{h}_k, \quad k = m_y, m_y - 1, \ldots, 1.$$

Using the fact that $t^*_{jk} = t_{kj}$ results in

$$(S + t_{kk}I)\vec{w}_k = \vec{h}_k - \sum_{j=k+1}^{m_y} t_{kj}\vec{w}_j, \quad k = m_y, \ldots, 1. \tag{5.113}$$

These upper triangular systems are easy to solve by back substitution. This yields $W^{(2)}$, which in turn yields $U^{(2)}$ via

$$U^{(2)} = D\left(\frac{1}{v}\right)QW^{(2)}R^*D\left(\frac{1}{w}\right).$$

There is a "real" version of this algorithm where Q and R are orthogonal, S is quasi-upper triangular and T is upper triangular [6], [7].

Example 5.6 Consider

$$-\Delta^{(2)}u(x,y) = f(x,y), \quad (x,y) \in S = (0,1) \times (0,1) \tag{5.114}$$
$$u(x,y) = 0, \quad (x,y) \in \partial S$$

where $f(x,y)$ is consistent with the solution $u(x,y) = xy\ell n(x)\ell n(y)$. This is a two-dimensional version of Example 4.8. The discrete system for this problem with weight function $v(x)w(y) = 1/\sqrt{\phi'(x)\phi'(y)}$ is given by

$$D\left(\frac{1}{x(1-x)}\right)\left[-\frac{1}{h^2}I^{(2)} + D\left(\frac{1}{4}\right)\right]D\left(\frac{1}{x(1-x)}\right)V^{(2)}$$
$$+V^{(2)}D\left(\frac{1}{y(1-y)}\right)\left[-\frac{1}{h^2}I^{(2)} + D\left(\frac{1}{4}\right)\right]D\left(\frac{1}{y(1-y)}\right)$$
$$= G^{(2)}$$

where

$$V^{(2)} = D\left(\sqrt{x(1-x)}\right)U^{(2)}D\left(\sqrt{y(1-y)}\right)$$

and

$$G^{(2)} = D\left(\sqrt{x(1-x)}\right)F^{(2)}D\left(\sqrt{y(1-y)}\right).$$

The solution $V^{(2)}$, after a matrix multiplication, gives the coefficients $u^s_{m_x,m_y}(x_i,y_j) = u^s_{ij}$ in the approximate Sinc-Galerkin solution (5.100). The parameters are $\alpha_s = \beta_s = \zeta_s = \eta_s = 1/2$ so $M_x = N_x = M_y = N_y$ and $h_s = \pi/\sqrt{M_x}$ ($d = \pi/2$). Since the nodes $x_i = e^{ih}/(1+e^{ih})$, and $y_j = e^{jh}/(1+e^{jh})$ are the same, the diagonal matrices in x and y are identical (same size and same entries). In Table 5.5 the maximum errors over the set of sinc gridpoints

$$S = \{x_i\}_{i=-M_x}^{N_x} \times \{y_j\}_{j=-M_y}^{N_y}$$

and the set of uniform gridpoints (stepsize $h_u = .01$)

$$U = \{w_i\}_{i=0}^{100} \times \{z_j\}_{j=0}^{100};$$
$$w_i = ih_u, \quad z_j = jh_u$$

are reported as

$$\|E_S^s(h_s)\| = \max_{\substack{-M_x \leq i \leq N_x \\ -M_y \leq j \leq N_y}} |u(x_i, y_j) - u_{ij}^s|$$

and

$$\|E_U^s(h_u)\| = \max_{\substack{0 \leq i \leq 100 \\ 0 \leq j \leq 100}} |u(w_i, z_j) - u_{m_x,m_y}^s(w_i, z_j)|,$$

respectively.

M_x	$h_s = \pi/\sqrt{M_x}$	$\|E_S^s(h_s)\|$	$\|E_U^s(h_u)\|$
2	2.221	$.168 \times 10^{-2}$	$.101 \times 10^{-1}$
4	1.571	$.665 \times 10^{-3}$	$.204 \times 10^{-2}$
8	1.111	$.110 \times 10^{-3}$	$.193 \times 10^{-3}$
16	0.785	$.586 \times 10^{-5}$	$.608 \times 10^{-5}$

Table 5.5 Errors for the two-dimensional Poisson problem (5.114). ■

Based on the numerical results shown in Example 5.6 and many other examples having similar results [13], the weight $v(x)w(y) = 1/\sqrt{\phi'(x)\phi'(y)}$ is recommended for the Poisson problem.

The development of the Sinc-Galerkin method for the Poisson problem in three dimensions mimics that in the previous section for two dimensions. The formulation as a Sylvester equation requires an extension to three dimensions and this in turn leads to an analogous formulation using the Kronecker product and the matrix (mat) operator described in Definition A.30. The general development can be found in [17].

Thus consider the Poisson problem in $\mathbb{R}^3$ given by

$$\begin{aligned} -\Delta^{(3)} u(x,y,z) &= f(x,y,z), \\ (x,y,z) \in S &= (0,1) \times (0,1) \times (0,1) \\ u(x,y,z) &= 0, \quad (x,y,z) \in \partial S. \end{aligned} \quad (5.115)$$

The approximate solution is given by

$$u^s_{m_x,m_y,m_z}(x,y,z) = \sum_{k=-M_z}^{N_z} \sum_{j=-M_y}^{N_y} \sum_{i=-M_x}^{N_x} u^s_{ijk} S_{ijk}(x,y,z) \quad (5.116)$$

where

$$S_{ijk}(x,y,z) = S(i,h_x) \circ \phi(x) S(j,h_y) \circ \phi(y) S(k,h_z) \circ \phi(z)$$

and

$$m_q = M_q + N_q + 1$$

for $q = x$, y, or z. The map for the basis functions is, as in the two-dimensional case, denoted by

$$\phi(z) = \ell n \left(\frac{z}{1-z}\right).$$

Wherever necessary, the independent variable will be denoted by subscripts ($\phi_q, q = x, y$, and z). The inner product is defined as

$$(f,g) = \int_0^1 \int_0^1 \int_0^1 f(x,y,z) g(x,y,z) w(x,y,z) dx dy dz$$

where

$$w(x,y,z) = \frac{1}{\sqrt{\phi'(x)\phi'(y)\phi'(z)}}.$$

This choice of weight is motivated by the discussion and numerical results in the previous example. Again the derivatives of u appearing in the orthogonalization of the residual for $-M_x \leq p \leq N_x$, $-M_y \leq q \leq N_y$, $-M_z \leq r \leq N_z$

$$(-\Delta^{(3)} u - f, S_{pqr}) = 0$$

are removed by applying Green's identity. Thus

$$\int_0^1 \int_0^1 \int_0^1 -\Delta^{(3)} u(x,y,z) S_{pqr}(x,y,z) w(x,y,z) dx dy dz$$

$$= \int_0^1 \int_0^1 \int_0^1 -u(x,y,z) \Delta^{(3)} \left(S_{pqr}(x,y,z) w(x,y,z)\right) dx dy dz$$

$$+ B_{T_3}$$

STEADY PROBLEMS

where

$$B_{T_3} = \iint_{\partial S}\left\{u(x,y,z)\frac{\partial}{\partial n}(S_{pqr}(x,y,z)w(x,y,z))\right.$$

$$\left. - S_{pqr}(x,y,z)w(x,y,z)\frac{\partial u}{\partial n}(x,y,z)\right\}dA$$

and n is the outward normal direction to S. Assuming $B_{T_3} = 0$ leads to the discretization

$$\mathcal{A}^{(2)}\mathrm{mat}(V^{(3)}) + \mathrm{mat}(V^{(3)})A_z = \mathrm{mat}(G^{(3)}) \tag{5.117}$$

where from (5.109)

$$\mathcal{A}^{(2)} = I_{m_y} \otimes A_x + A_y \otimes I_{m_x}$$

and A_z is defined by (5.89) with x replaced by z. The unknowns are given by

$$\mathrm{mat}(V^{(3)}) = D\left((\phi'_y)^{-1/2}\right)$$
$$\otimes D\left((\phi'_x)^{-1/2}\right)\mathrm{mat}(U^{(3)})D\left((\phi'_z)^{-1/2}\right)$$
$$\equiv D_{yx}\mathrm{mat}(U^{(3)})D_z$$

and the discretization of the forcing term by

$$\mathrm{mat}(G^{(3)}) = D\left((\phi'_y)^{-1/2}\right)$$
$$\otimes D\left((\phi'_x)^{-1/2}\right)\mathrm{mat}(F^{(3)})D\left((\phi'_z)^{-1/2}\right)$$
$$\equiv D_{yx}\mathrm{mat}(F^{(3)})D_z.$$

Lastly the matrix operator mat from Definition A.30 is given as

$$\mathrm{mat}(U^{(3)}) = \mathrm{mat}((u^s_{ijk}))$$
$$= \left[\mathrm{co}(u^s_{i,j,-M_z}), \mathrm{co}(u^s_{i,j,-M_z+1}), \ldots, \mathrm{co}(u^s_{i,j,N_z})\right]$$

and similarly for $\mathrm{mat}(F^{(3)})$. Thus $\mathrm{mat}(U^{(3)})$ is an $m_x m_y \times m_z$ matrix, where each column is the concatenation of the x and y sinc gridpoints in natural order and the columns of $\text{mat}(U^{(3)})$ are ordered from smallest to largest z value.

As in the two-dimensional case, the application of Theorem 4.11 to each of the three one-dimensional problems in the x, y, and z variables leads to the following requirements. For $q = x$, y or z, $f/\sqrt{\phi'_q} \in B(D_E)$, and $uF_q \in B(D_E)$, where

$$F_q = \sqrt{\phi'_q}\left(1/\phi'_q\right)'', \quad (\phi'_q)^{3/2}, \quad \phi''_q/\sqrt{\phi'_q}, \quad \left(1/\sqrt{\phi'_q}\right)''.$$

Furthermore, assume

$$|u(x,y,z)| \le Cx^{\alpha_s+1/2}(1-x)^{\beta_s+1/2}y^{\zeta_s+1/2}$$
$$(1-y)^{\eta_s+1/2}z^{\mu_s+1/2}(1-z)^{\nu_s+1/2}$$

and make the selections (5.95), (5.96), and (5.97) as before (using α_s, β_s, ζ_s, η_s, and h_s), while supplementing with

$$h_s = h_z = h_y = h_x = \left(\frac{\pi d}{\alpha_s M_x}\right)^{1/2},$$

$$M_z = \left[\left|\frac{\alpha_s}{\mu_s} M_x + 1\right|\right] \tag{5.118}$$

and

$$N_z = \left[\left|\frac{\alpha_s}{\nu_s} M_x + 1\right|\right]. \tag{5.119}$$

These selections lead to

$$\|u - u^s_{m_x,m_y,m_z}\|_\infty$$
$$\le K_s M_x^2 \exp\left(-(\pi d \alpha_s M_x)^{1/2}\right). \tag{5.120}$$

Concatenating (5.117) results in the Kronecker sum form given by

$$(I_{m_z} \otimes \mathcal{A}^{(2)} + A_z \otimes I_{m_y m_x})\text{co}(V^{(3)}) = \text{co}(G^{(3)}) \tag{5.121}$$

where from Definitions A.29 and A.30 the equalities

$$\begin{aligned}\text{co}(\text{mat}(V^{(3)})) &= \text{co}(V^{(3)}) \\ &= [D((\phi'_z)^{-1/2}) \otimes D((\phi'_y)^{-1/2}) \\ &\quad \otimes D((\phi'_x)^{-1/2})]\text{co}(U^{(3)}) \\ &\equiv D_{zyx}\text{co}(U^{(3)})\end{aligned} \tag{5.122}$$

STEADY PROBLEMS   213

and
$$\text{co}(\text{mat}(G^{(3)})) = \text{co}(G^{(3)}) \qquad (5.123)$$
$$= D_{zyx}\text{co}(F^{(3)})$$

have been used. Letting
$$\mathcal{A}^{(3)} \equiv I_{m_z} \otimes \mathcal{A}^{(2)} + A_z \otimes I_{m_y m_x} \qquad (5.124)$$
$$= I_{m_z} \otimes I_{m_y} \otimes A_x + I_{m_z} \otimes A_y \otimes I_{m_x} + A_z \otimes I_{m_y} \otimes I_{m_x}$$

then (5.121) may be written
$$\mathcal{A}^{(3)}\text{co}(V^{(3)}) = \text{co}(G^{(3)}). \qquad (5.125)$$

The solution of (5.117) (or equivalently (5.125)) is given in what follows. Let P and Q be the orthogonal diagonalizers of the matrices A_x and A_y in (5.107) and (5.108), respectively. Let R denote the orthogonal diagonalizer of A_z so that
$$R^T A_z R = \Lambda_z. \qquad (5.126)$$

Let the orthogonal matrix Q_{zyx} be given by
$$Q_{zyx} \equiv R \otimes Q \otimes P$$

and define
$$\text{co}(W^{(3)}) = Q_{zyx}^T \text{co}(V^{(3)})$$
$$\text{co}(H^{(3)}) = Q_{zyx}^T \text{co}(G^{(3)}).$$

Then (5.125) becomes
$$(I_{m_z} \otimes I_{m_y} \otimes \Lambda_x + I_{m_z} \otimes \Lambda_y \otimes I_{m_x} + \Lambda_z \otimes I_{m_y} \otimes I_{m_x})\text{co}(W^{(3)}) = \text{co}(H^{(3)})$$

whose solution is
$$w_{ijk} = \frac{h_{ijk}}{(\lambda_x)_i + (\lambda_y)_j + (\lambda_z)_k}, \quad \begin{array}{c} -M_x \leq i \leq N_x,\ -M_y \leq j \leq N_y \\ -M_z \leq k \leq N_z \end{array}.$$

Then $U^{(3)}$ is recovered from $W^{(3)}$ by using

$$\begin{aligned}\text{co}(U^{(3)}) &= D_{zyx}^{-1} Q_{zyx} \text{co}(W^{(3)}) \\ &= [D((\phi_z')^{1/2})R \otimes D((\phi_y')^{1/2})Q \\ &\quad \otimes D((\phi_x')^{1/2})P]\text{co}(W^{(3)}).\end{aligned}$$

The block structure of the matrix $\mathcal{A}^{(3)}$ in (5.124) is illustrated in Figure 5.3.

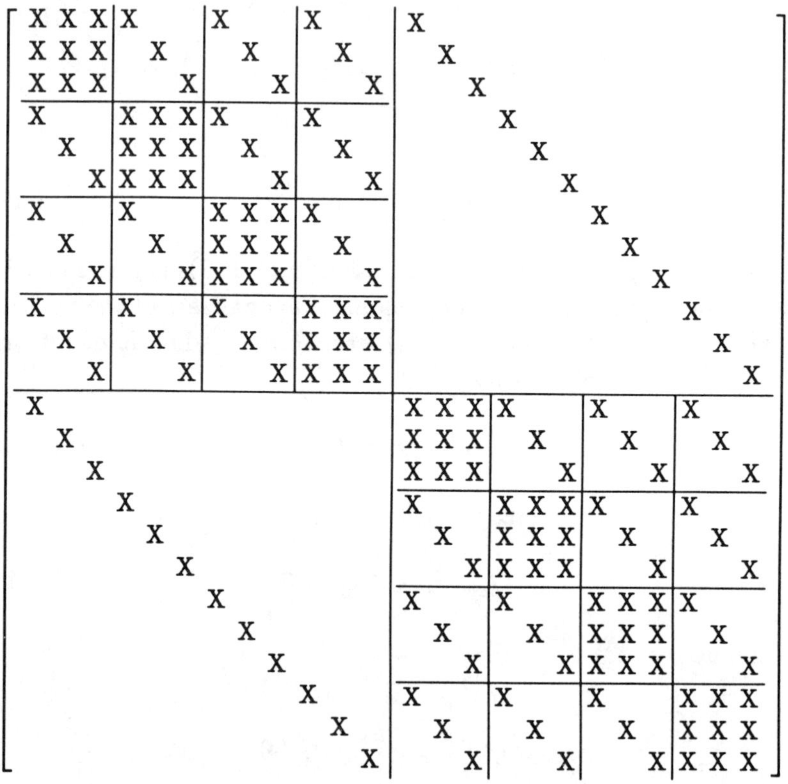

$$\begin{bmatrix} X\ X\ X & X & & X & & X & & X & & & & & & \\ X\ X\ X & & X & & X & & X & & X & & & & & \\ X\ X\ X & & X & & X & & X & & X & & & & & \\ X & X\ X\ X & X & X & & & & X & & & & & & \\ X & X\ X\ X & X & & X & & & X & & & & & \\ & X & X\ X\ X & & X & & X & & X & & & & & \\ X & X & X\ X\ X & X & & & & & X & & & & \\ X & X & X\ X\ X & X & & & & X & & & & \\ X & X & X\ X\ X & X & & & & X & & & \\ X & X & X & X\ X\ X & & & & & X & & \\ X & X & X & X\ X\ X & & & & & X & \\ X & X & X\ X\ X & & & & & & X \\ X & & & & & X\ X\ X & X & X & X & & & \\ X & & & & X\ X\ X & X & X & X & & & \\ X & & & & X\ X\ X & X & X & X & & \\ & X & & & & X & X\ X\ X & X & X & & \\ & X & & & X & X\ X\ X & X & X & & \\ & X & & & X\ X\ X & X & X & \\ & & X & & X & X & X\ X\ X & X \\ & & X & & X & X & X\ X\ X & X \\ & & X & & X & X & X\ X\ X & X \\ & & & X & X & X & X & X\ X\ X \\ & & & X & X & X & X & X\ X\ X \\ & & & X & X & X & X & X\ X\ X \end{bmatrix}$$

Figure 5.3 Nonzero elements of $\mathcal{A}^{(3)}$ for $m_x = 3$, $m_y = 4$, $m_z = 2$.

The calculation of $U^{(3)}$ in this recovery is straightforward. To see this, let

$$R_w = D((\phi_z')^{1/2})R, \quad Q_w = D((\phi_y')^{1/2})Q, \quad P_w = D((\phi_x')^{1/2})P$$

where $(P_w)_{jk} = p^w_{jk}$ and similarly for Q_w and R_w. Then the ijk-th component of $U^{(3)}$ is recovered from $(W^{(3)})_{ijk} = w_{ijk}$ by

$$(U^{(3)})_{ijk} = \sum_{t=1}^{m_z} r^w_{kt} \sum_{s=1}^{m_y} q^w_{js} \sum_{r=1}^{m_x} p^w_{ir} w_{rst}.$$

Example 5.7 Consider

$$\begin{aligned}
-\Delta^{(3)} u(x,y,z) &= f(x,y,z), \\
(x,y,z) \in S &= (0,1) \times (0,1) \times (0,1) \quad (5.127) \\
u(x,y,z) &= 0, \quad (x,y,z) \in \partial S
\end{aligned}$$

where $u(x,y,z) = -10x(1-x)^{3/2} y^{4/3}(1-y)^2 z^3 (1-z)^{5/3}$. The solution exhibits singularities along $x = 1$, $y = 0$, and $z = 1$. The parameters are given by $\alpha_s = 1/2$, $\beta_s = 1$, $\zeta_s = 5/6$, $\eta_s = 3/2$, $\mu_s = 5/2$, and $\eta_s = 7/6$ so the stepsize $h_s = \pi/\sqrt{M_x}$. The values of M_x, N_x, M_y, N_y, M_z, and N_z are reported in Table 5.6, along with the error at the sinc gridpoints denoted $\|E^s_S(h_s)\|$.

M_x	N_x	M_y	N_y	M_z	N_z	h_s	$\|E^s_S(h_s)\|$
4	2	3	2	1	2	1.571	$.298 \times 10^{-3}$
8	4	5	3	2	4	1.111	$.390 \times 10^{-4}$
16	8	10	6	4	7	0.785	$.583 \times 10^{-5}$

Table 5.6 Errors for the three-dimensional Poisson problem in (5.127).

The discrete system corresponding to (5.125) has components (for $q = x, y, z$)

$$A_q = D(\phi'_q) \left[-\frac{1}{h_s^2} I^{(2)} + D\left(\frac{1}{4}\right) \right] D(\phi'_q)$$

where
$$\phi'_q(q) = \frac{1}{q(1-q)}.$$

This system is solved for $V^{(3)}$, from which the coefficients u^s_{ijk} for the Sinc-Galerkin solution (5.116) are obtained from (5.122). ∎

The various approaches to the solution of (5.84) (or equivalently (5.101)) and their three-dimensional analogues (5.117) and (5.125) have been algorithmically described. However, certain assumptions have been made that will be addressed here. The invertibility of $\mathcal{A}^{(2)}$ in (5.101) and $\mathcal{A}^{(3)}$ in (5.125) is equivalent to the unique solvability of (5.84) and (5.117), respectively. The discussion following Theorem 4.19 using the properties of $I^{(2)}$ and $I^{(1)}$ given in Theorems 4.18 and 4.19, respectively, sheds some light on the invertibility issue. More importantly, the spectral analysis below yields bounds on the conditioning of the coefficient matrices. The results will be stated for an arbitrary finite interval (a_q, b_q), $q = 1$ or 2.

The solvability of the matrix system

$$\mathcal{A}^{(2)}\text{co}(V^{(2)}) = \text{co}(G^{(2)}) \tag{5.128}$$

where $\mathcal{A}^{(2)}$ is given in (5.109) is equivalent to the statement that no eigenvalue of A_x is the negative of any eigenvalue of A_y (Theorem A.27). From Theorem 4.18, $I^{(2)}$ is a symmetric, negative definite matrix. Thus the matrix

$$A\left(\frac{1}{\sqrt{\phi'}}\right) = -\frac{1}{h^2} I^{(2)} + D\left(\frac{1}{4}\right)$$

is positive definite and symmetric. Now on any finite interval (a, b), $\phi(x) = \ell n \left(\frac{x-a}{b-x}\right)$ and

$$\phi'(x) = \frac{b-a}{(x-a)(b-x)}. \tag{5.129}$$

So $\phi'(x) > 0$ on (a, b) and thus $D(\phi'_x)$ is a symmetric, positive definite matrix (the same holds for $D(\phi'_y)$). Hence A_x and A_y are symmetric, positive definite matrices. This in turn says that if the eigenvalues

STEADY PROBLEMS 217

of A_x and A_y are denoted $(\lambda_x)_{-M_x} \leq \cdots \leq (\lambda_x)_{N_x}$ and $(\lambda_y)_{-M_y} \leq \cdots \leq (\lambda_y)_{N_y}$, respectively, then $(\lambda_x)_{-M_x} > 0$ and $(\lambda_y)_{-M_y} > 0$. Thus, by Theorem A.27, the spectrum of $\mathcal{A}^{(2)}$ is

$$\sigma(\mathcal{A}^{(2)}) = \{(\lambda_x)_i + (\lambda_y)_j\}_{\substack{-M_x \leq i \leq N_x \\ -M_y \leq j \leq N_y}}$$

which does not contain zero so that $\mathcal{A}^{(2)}$ is invertible. A little more careful analysis yields the following lower bound for the eigenvalues of A_x [4].

Theorem 5.8 *If $(\lambda_x)_{-M_x}$ is the smallest eigenvalue of the matrix A_x, then*

$$(\lambda_x)_{-M_x} \geq \frac{4}{(b-a)^2}.$$

Proof Consider A_x and recall from Theorem 4.18 that $\sigma(-I^{(2)}) \subset (0, \pi^2]$. Since $h > 0$, $\sigma(\mathcal{A}(1/\sqrt{\phi'})) \subset \left(\frac{1}{4}, \frac{\pi^2}{h^2} + \frac{1}{4}\right]$. Now from (5.129)

$$\phi'(x) \geq \frac{4}{b-a}.$$

Thus from the Courant–Fischer Minimax Theorem A.17, with $\vec{x} \in \mathbb{C}^n$ and $\vec{y} = D(\phi'_x)\vec{x}$

$$\begin{aligned}
(\lambda_x)_{-M_x} &= \min_{\vec{x} \neq \vec{0}} \left(\frac{\vec{x}^* A_x \vec{x}}{\vec{x}^* \vec{x}}\right) \\
&= \min_{\vec{x} \neq \vec{0}} \left(\frac{\vec{x}^* D(\phi'_x) \mathcal{A}(1/\sqrt{\phi'}) D(\phi'_x) \vec{x}}{\vec{x}^* \vec{x}}\right) \\
&= \min_{\vec{x} \neq \vec{0}} \left(\frac{\vec{x}^* D(\phi'_x) \mathcal{A}(1/\sqrt{\phi'}) D(\phi'_x) \vec{x}}{x^* D(\phi'_x) D(\phi'_x) \vec{x}} \cdot \frac{\vec{x}^* D(\phi'_x) D(\phi'_x) \vec{x}}{\vec{x}^* \vec{x}}\right) \\
&\geq \min_{\vec{y} \neq \vec{0}} \left(\frac{\vec{y}^* \mathcal{A}(1/\sqrt{\phi'}) \vec{y}}{\vec{y}^* \vec{y}}\right) \min_{\vec{x} \neq \vec{0}} \left(\frac{\vec{x}^* (D(\phi'_x))^2 \vec{x}}{\vec{x}^* x}\right) \\
&\geq \left(\frac{1}{4}\right)\left(\frac{4}{b-a}\right)^2 \\
&= \frac{4}{(b-a)^2}.
\end{aligned}$$

∎

Combining Theorem 5.8 with Theorem A.27 then gives the following corollary.

Corollary 5.9 *The matrix*

$$\mathcal{A}^{(2)} = I_{m_y} \otimes A_x + A_y \otimes I_{m_x}$$

is a positive definite matrix with eigenvalues λ bounded below by

$$\lambda \geq 4 \left(\frac{1}{(b_1 - a_1)^2} + \frac{1}{(b_2 - a_2)^2} \right)$$

where the original problem (5.79) is defined on $(a_1, b_1) \times (a_2, b_2)$.

■

Exercises

Exercise 5.1 Take $w(x) = (\phi'(x))^{-r}$ in (5.10) to obtain (5.11). If $p(x) \equiv 0$ and $r = 1/2$ then this is the symmetric system in (4.61). For the purposes of this problem, denote the coefficients in the system (5.11) by $\vec{u}_g$. If $r = 1$ (for any $p(x)$) this system is the Sinc-Galerkin system

$$\left\{ \mathcal{A}\left(\frac{1}{\phi'}\right) - \frac{1}{h} I^{(1)} D\left(\frac{p}{\phi'}\right) + D\left(\frac{-1}{\phi'}\left(\frac{p}{\phi'}\right)'\right) \right.$$
$$\left. + D\left(\frac{q}{(\phi')^2}\right) \right\} \vec{u}_g = D\left(\frac{1}{(\phi')^2}\right) \vec{f} \quad (5.130)$$

where, from (5.12), the matrix $\mathcal{A}(1/\phi')$ is defined by

$$\mathcal{A}\left(\frac{1}{\phi'}\right) \equiv \frac{-1}{h^2} I^{(2)} + \frac{1}{h} I^{(1)} D\left(\frac{\phi''}{(\phi')^2}\right) + D\left(\frac{-1}{\phi'}\left(\frac{1}{\phi'}\right)''\right).$$

To continue the discussion following (5.29) on the analogy between the Sinc-Galerkin and sinc-collocation schemes, recall that the discrete system for the collocation scheme (5.24), with $\ell = 0$, reads

$$C(0)\vec{u}_c = \left\{ \frac{-1}{h^2} I^{(2)} - \frac{1}{h} D\left(\frac{-\phi''}{(\phi')^2} + \frac{p}{\phi'}\right) I^{(1)} + D\left(\frac{q}{(\phi')^2}\right) \right\} \vec{u}_c$$
$$= D\left(\frac{1}{(\phi')^2}\right) \vec{f} \quad (5.131)$$

STEADY PROBLEMS 219

where $\vec{u}_c$ are the coefficients in (5.25).

Assume that the coefficients $\vec{u}_c$ and $\vec{u}_g$ are computed to within the error of the method and set $\vec{u}_c = \vec{u}_g = \vec{u}$. Subtract the collocation scheme from the Galerkin scheme to find

$$\left\{ I^{(1)} D\left(\frac{\phi''}{(\phi')^2} - \frac{p}{\phi'}\right) - D\left(\frac{\phi''}{(\phi')^2} - \frac{p}{\phi'}\right) I^{(1)} \right\} \vec{u}$$

$$= hD\left(\frac{1}{\phi'}\left(\frac{1}{\phi'}\right)'' + \frac{1}{\phi'}\left(\frac{p}{\phi'}\right)'\right) \vec{u} \quad (5.132)$$

$$= hD\left(\frac{1}{\phi'}\left[\left(\frac{1}{\phi'}\right)'' + \left(\frac{p}{\phi'}\right)'\right]\right) \vec{u}.$$

Multiply this result by -1, note that $\phi''/(\phi')^2 = -(1/\phi')'$ and recall from (5.16) that $\xi_p = -\phi''/(\phi')^2 + p/\phi'$ to arrive at the identity

$$\left(I^{(1)} D\left(\xi_p\right) - D\left(\xi_p\right) I^{(1)} \right) \vec{u} = -hD\left(\frac{\xi_p'}{\phi'}\right) \vec{u}. \quad (5.133)$$

Solve this equation for $I^{(1)} D(\xi_p)$. Substitute this result in the Sinc-Galerkin scheme (5.130) to arrive at the sinc-collocation scheme (5.131).

The commutator of the matrix $I^{(1)}$ with a diagonal matrix defined by the function ξ_p in (5.133) yields (approximately) a weighted derivative of the function. Replacing ξ_p by a differentiable function τ, one is led to the following question. For what class of functions does the (approximate) identity

$$\left(I^{(1)} D\left(\tau\right) - D\left(\tau\right) I^{(1)} \right) \vec{u} = -hD\left(\frac{\tau'}{\phi'}\right) \vec{u}$$

hold? That is, in maximum norm, does the error satisfy

$$\max_{-M \leq j \leq N} \left| \left\{ I^{(1)} D\left(\tau\right) - D\left(\tau\right) I^{(1)} + hD\left(\frac{\tau'}{\phi'}\right) \right\} \vec{u} \right| \leq C e^{-\kappa \sqrt{M}},$$

where C and κ are positive constants? The matrices and vectors are $(M+N+1) \times (M+N+1)$ and $(M+N+1) \times 1$, respectively. The function $\tau(x_j)$ is evaluated at the nodes $x_j = \psi(jh)$ and N and h are defined by Theorem 4.7. If $\tau(x)$ is a constant then all is copacetic!

Exercise 5.2 A case of interest is the self-adjoint problem (5.1), i.e., $p(x) \equiv 0$. If $\phi(x) = ln((x-a)/(b-x))$ in (5.132) and the assumption $\vec{u}_c = \vec{u}_g = \vec{u}$ remains in force, derive the approximate identity

$$\left(I^{(1)}D(x) - D(x)I^{(1)}\right)\vec{u} = -hD\left(\frac{1}{\phi'}\right)\vec{u}.$$

In the case that $\phi(x) = ln(x)$ and $p(x) \equiv 0$, show that the Sinc-Galerkin system (5.130) and the sinc-collocation system (5.131) are identical.

Exercise 5.3 Recompute the numerical solution to the boundary value problem in Example 5.1 using the system in (5.131). Notice that the collocation system of the present exercise, which is given by

$$\left\{\frac{-1}{h^2}I^{(2)} - \frac{1}{h}D\left(\frac{5-11x}{6}\right)I^{(1)} + D\left((1-x)^2\right)\right\}\vec{u}$$
$$= D\left((x(1-x))^2\right)\vec{f},$$

is different from the one used in that example. The numerical results are almost identical to the results in Table 5.1 (Example 5.1) under the heading $\|E_S^G(h)\|$.

Exercise 5.4 Bessel's differential equation of order ν in self-adjoint form is

$$-\frac{d}{dx}\left(x\frac{dy}{dx}\right) + \frac{\nu^2}{x}y = \lambda xy, \quad 0 < x < 1$$
$$y(0) = y(1) = 0.$$

Make the change of variable

$$y(x) = x^{-1/2}u(x)$$

to write the equation in the normal form (given in (5.33))

$$-\frac{d^2u}{dx^2} + \left(\frac{4\nu^2 - 1}{4x^2}\right)u(x) = \lambda u(x), \quad 0 < x < 1$$
$$u(0) = u(1) = 0.$$

A sinc discretization for the self-adjoint form is available from the system in (5.77) with $c(x) = x$. The discretization for the normal

STEADY PROBLEMS   221

form is obtained from (5.40). Inspect these two systems and, recalling the skew-symmetry of the matrix $I^{(1)}$, give at least one reason why the discretization for the normal form is preferable over the system obtained from (5.77).

Exercise 5.5 The radial Schrödinger equation with a Woods–Saxon potential takes the form

$$-\frac{d^2u}{dx^2} + u(x) = \lambda \left(1 + e^{(x-r)/\epsilon}\right)^{-1} u(x), \quad 0 < x < \infty$$

$$u(0) = \lim_{x\to\infty} u(x) = 0$$

so that $q(x) \equiv 1$ and

$$\rho(x) = \left(1 + e^{(x-r)/\epsilon}\right)^{-1}$$

in (5.31). This system is singular due to the semi-infinite interval. Take $r = 5.086855$ and $\epsilon = .929853$ and compute the solution using the $\ell n(\sinh(x))$ map with $d = \pi/2$. The first two eigenvalues are $\lambda = 1.424333$ and $\lambda = 2.444704$. The function $\rho(x)$ has a pole at $\tan^{-1}(\epsilon\pi/r)$. Discuss why it is necessary to choose $d \approx \pi/6$ if the map $\ell n(x)$ is used for the discretization [19].

Exercise 5.6 Use the system in (5.73) to compute the coefficients of c_{m_c} in (5.70) for the problem (5.78). By interpolation, the coefficients c_j approximate some solution $c_\delta(x)$ of (5.78) evaluated at x_j. Put $\delta = 0$ and calculate each of the errors

$$\max_{-M_c \leq j \leq M_c} |c_0(x_j) - c_j|$$

and

$$\max_{\substack{-M_c \leq j \leq M_c \\ j \neq 0, \pm 1}} |c_0(x_j) - c_j|$$

for $M_c = 4, 8,$ and 16. The former appears stuck while the latter is decreasing. This is what was meant by the statement following (5.78) indicating that the error is concentrated around the point $x = 1/2$. The node $x_0 = 1/2$ is the midpoint of the interval $(0,1)$ and the two points $x_{\pm 1}$ are the nodes to the immediate right and left of $x = 1/2$, respectively.

Exercise 5.7 Consider the Helmholtz equation

$$-\Delta^{(2)}u(x,y) = \lambda u(x,y), \quad (x,y) \in S = (0,1) \times (0,1) \tag{5.134}$$

$$u(x,y) = 0, \quad (x,y) \in \partial S$$

which has the eigenvalues $\lambda_{nm} = (n^2 + m^2)\pi^2$, for all positive integers m and n. Use the left-hand side of the system in (5.90) to find the discrete system

$$A_x V^{(2)} + V^{(2)} A_y = \mu V^{(2)} \tag{5.135}$$

for the approximation on the eigenvalues of (5.134). The equivalent matrix form of (5.135) is

$$\mathcal{A}^{(2)}\mathrm{co}(V^{(2)}) = \mu\,\mathrm{co}(V^{(2)})$$

where

$$\mathcal{A}^{(2)} = I_{m_y} \otimes A_x + A_y \otimes I_{m_x}$$

and

$$A_x = D\left(\frac{1}{x(1-x)}\right)\left[-\frac{1}{h^2}I^{(2)} + D\left(\frac{1}{4}\right)\right]D\left(\frac{1}{x(1-x)}\right)$$

(similarly for A_y). Calculate the smallest eigenvalue of $\mathcal{A}^{(2)}$ (identified as $(\mu_{xy})_{-M_x,-M_y}$). The parameters used are $\alpha_s = \beta_s = \zeta_s = \eta_s = 1/2$, so $M_x = N_x = M_y = N_y$ and $h_s = \pi/\sqrt{M_x}$ ($d = \pi/2$). You should find that $(\mu_{xy})_{-M_x,-M_y}$ is approximating the true smallest eigenvalue of (5.134), which is $\mu = 2\pi^2 \approx 19.73921$. Show that the matrix A_x is the matrix in (5.40) for the approximation of the eigenvalues of the Fourier problem

$$-u''(x) = n^2\pi^2 u(x), \quad x \in (0,1)$$

$$u(0) = u(1) = 0$$

for all positive integers n. Calculate the smallest eigenvalue of A_x. The smallest eigenvalue of A_y is the same number. The sum of these two numbers is $(\mu_{xy})_{-M_x,-M_y}$. This follows from Theorem A.27.

References

[1] H. T. Banks and K. Kunisch, *Estimation Techniques for Distributed Parameter Systems*, Birkhäuser, Boston, 1989.

[2] B. Bialecki, "Sinc-Collocation Methods for Two-Point Boundary Value Problems," *IMA J. Numer. Anal.*, 11 (1991), pages 357–375.

[3] K. L. Bowers and J. Lund, "Numerical Solution of Singular Poisson Problems via the Sinc-Galerkin Method," *SIAM J. Numer. Anal.*, 24 (1987), pages 36–51.

[4] R. R. Doyle, "Extensions to the Development of the Sinc-Galerkin Method for Parabolic Problems," Ph.D. Thesis, Montana State University, Bozeman, MT, 1990.

[5] N. Eggert, M. Jarratt, and J. Lund, "Sinc Function Computation of the Eigenvalues of Sturm–Liouville Problems," *J. Comput. Phys.*, 69 (1987), pages 209–229.

[6] G. H. Golub, S. Nash, and C. Van Loan, "A Hessenberg–Schur Method for the Problem $AX+XB = C$," *IEEE Trans. Automat. Control*, 24 (1979), pages 909–913.

[7] G. H. Golub and C. F. Van Loan, *Matrix Computations*, Second Ed., Johns Hopkins University Press, Baltimore, 1989.

[8] M. Jarratt, "Approximation of Eigenvalues of Sturm–Liouville Differential Equations by the Sinc-Collocation Method," Ph. D. Thesis, Montana State University, Bozeman, MT, 1987.

[9] M. Jarratt, "Eigenvalue Approximations on the Entire Real Line," in *Computation and Control*, Proc. Bozeman Conf. 1988, Progress in Systems and Control Theory, Vol. 1, Birkhäuser, Boston, 1989, pages 133–144.

[10] M. Jarratt, "Eigenvalue Approximations for Numerical Observability Problems," in *Computation and Control II*, Proc. Bozeman Conf. 1990, Progress in Systems and Control Theory, Vol. 11, Birkhäuser, Boston, 1991, pages 173–185.

[11] M. Jarratt, J. Lund, and K. L. Bowers, "Galerkin Schemes and the Sinc-Galerkin Method for Singular Sturm–Liouville Problems," *J. Comput. Phys.*, 89 (1990), pages 41–62.

[12] J. Lund, "Sinc Approximation Method for Coefficient Identification in Parabolic Systems," in *Robust Control of Linear Systems and Nonlinear Control, Vol. II*, Proc. MTNS 1989, Progress in Systems and Control Theory, Vol. 4, Birkhäuser, Boston, 1990, pages 507–514.

[13] J. Lund, K. L. Bowers, and K. M. McArthur, "Symmetrization of the Sinc-Galerkin Method with Block Techniques for Elliptic Equations," *IMA J. Numer. Anal.*, 9 (1989), pages 29–46.

[14] J. Lund and B. V. Riley, "A Sinc-Collocation Method for the Computation of the Eigenvalues of the Radial Schrödinger Equation," *IMA J. Numer. Anal.*, 4 (1984), pages 83–98.

[15] J. Lund and C. R. Vogel, "A Fully-Galerkin Method for the Numerical Solution of an Inverse Problem in a Parabolic Partial Differential Equation," *Inverse Problems*, 6 (1990), pages 205–217.

[16] K. M. McArthur, "A Collocative Variation of the Sinc-Galerkin Method for Second Order Boundary Value Problems," in *Computation and Control*, Proc. Bozeman Conf. 1988, Progress in Systems and Control Theory, Vol. 1, Birkhäuser, Boston, 1989, pages 253–261.

[17] K. M. McArthur, K. L. Bowers, and J. Lund, "The Sinc Method in Multiple Space Dimensions: Model Problems," *Numer. Math.*, 56 (1990), pages 789–816.

[18] K. M. McArthur, R. C. Smith, J. Lund, and K. L. Bowers, "The Sinc-Galerkin Method for Parameter Dependent Self-Adjoint Problems," accepted by *Appl. Math. Comput.*

[19] B. V. Riley, "Galerkin Schemes for Elliptic Boundary Value Problems," Ph. D. Thesis, Montana State University, Bozeman, MT, 1982.

[20] R. C. Smith, "Numerical Solution of Fourth-Order Time-Dependent Problems with Applications to Parameter Identification," Ph. D. Thesis, Montana State University, Bozeman, MT, 1990.

[21] R. C. Smith and K. L. Bowers, "A Fully Galerkin Method for the Recovery of Stiffness and Damping Parameters in Euler-Bernoulli Beam Models," in *Computation and Control II*, Proc. Bozeman Conf. 1990, Progress in Systems and Control Theory, Vol. 11, Birkhäuser, Boston, 1991, pages 289–306.

[22] R. C. Smith, K. L. Bowers, and C. R. Vogel, "Numerical Recovery of Material Parameters in Euler-Bernoulli Beam Models," accepted by *J. Math. Systems, Estimation, and Control.*

[23] J. L. Schwing, "Numerical Solutions of Problems in Potential Theory," Ph.D. Thesis, University of Utah, Salt Lake City, UT, 1976.

[24] F. Stenger, "Numerical Methods Based on Whittaker Cardinal, or Sinc Functions," *SIAM Rev.*, 23 (1981), pages 165–224.

Chapter 6

Time-Dependent Problems

This chapter assembles the results in Chapters 4 and 5 to define, via a product formulation, a fully Sinc-Galerkin method for parabolic time-dependent problems. This fully Sinc-Galerkin method is described in Section 6.1, where the application is to the heat equation in one and two space dimensions. The unique solvability of the discrete system is also addressed. The method of handling nonhomogeneous Dirichlet boundary conditions and nonhomogeneous initial conditions is treated the same way it was in Chapter 4. The inclusion of radiation boundary conditions, by way of the bordered matrices met in Section 4.4, is also sorted out in Section 6.2. Indeed, the (linear) convection dominated transport problem closing Section 6.2 is the vehicle used to advertise a future direction in the final Section 6.3. This section addresses a computational scheme to iteratively solve the fully Sinc-Galerkin discretization for Burgers' equation.

6.1 The Heat Equation

The fully Sinc-Galerkin method described in this section, at one level, simply consists of the assembly of the discrete Sinc-Galerkin expansion of a second-order boundary value problem in the spatial domain with a Sinc-Galerkin discretization in the temporal domain for a first-order problem. By referring to the first-order material in Section 4.3 (in particular the system in (4.78)) and the material in Section 5.4 (in particular the system in (5.84)), one could, with a simple wave of the hand, amalgamate the two systems to arrive at a system for the heat problem. All of this actually works out and is not such a bad line of retrospective thinking. But for the purpose of clearly defining the system's roots, as well as analogizing with the earlier development, the story begins with the residual. The basis functions, trial functions, and the inner product are direct extensions of those described in Section 4.2. The two formulations given are, as was the case for the two-dimensional Poisson problem, tied together by the material on Kronecker sums given in Appendix A.3.

The heat problem is defined by the system

$$\begin{aligned} P^{(2)}u(x,t) &\equiv \frac{\partial u(x,t)}{\partial t} - \frac{\partial^2 u(x,t)}{\partial x^2} = f(x,t) \\ u(0,t) &= u(1,t) = 0, \quad t > 0 \\ u(x,0) &= 0, \quad 0 < x < 1. \end{aligned} \quad (6.1)$$

The approximate solution to (6.1) is defined by

$$u_{m_x,m_t}(x,t) = \sum_{j=-M_t}^{N_t} \sum_{i=-M_x}^{N_x} u_{ij} S_{ij}(x,t) \quad (6.2)$$

where $m_x = M_x + N_x + 1$ and $m_t = M_t + N_t + 1$. The basis functions $\{S_{ij}(x,t)\}$ for $-M_x \le i \le N_x$, $-M_t \le j \le N_t$ are given as the product of basis functions for the appropriate one-dimensional problem. For this chapter their definition is recalled from (4.6) so that

$$\begin{aligned} S_{ij}(x,t) &\equiv S_i(x) S_j^*(t) \\ &\equiv S(i, h_x) \circ \phi(x) S(j, h_t) \circ \Upsilon(t) \end{aligned} \quad (6.3)$$

where the conformal map in the spatial domain is given by

$$\phi(z) = \ell n\left(\frac{z}{1-z}\right). \tag{6.4}$$

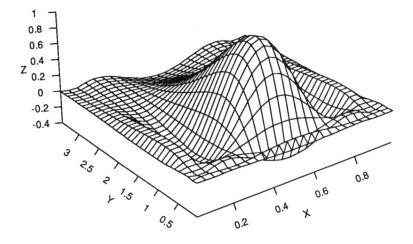

Figure 6.1 The mapped sinc basis in (6.3) on $S = (0,1) \times (0,\infty)$ for $i = j = 0$ and $h = \pi/4$.

∎

The map in the temporal domain will be denoted by Υ throughout this chapter. The matrix resulting from this discretization in the temporal domain is significantly different from the matrix arising from the discretization in the spatial domain, and the symbol Υ is a more emphatic delineation of the matrix system than is ϕ_t. The examples in this chapter are limited to the map (see [5], [10], and [11])

$$\Upsilon(t) = \ell n(t). \tag{6.5}$$

This map has also been used in fourth-order time-dependent partial differential equations [13]. However, the entire development presented here applies to the alternative selection $\Upsilon(t) = \ell n(\sinh(t))$ with minor modifications [2]. The few places where a change is needed will be explicitly stated. For this reason, the general discussion will build the matrices of the discrete system for (6.1) with generic Υ.

As is now the convention, orthogonalize the residual by first considering the inner products in

$$\left(P^{(2)}u - f, S_{k\ell}\right) = 0,$$

$-M_x \leq k \leq N_x$, and $-M_t \leq \ell \leq N_t$, where the inner product is defined by

$$(f, g) = \int_0^\infty \int_0^1 f(x,t)g(x,t)w(x)\tau(t)\,dx\,dt$$

and $w(x)\tau(t)$ is a product weight function. For reasons that are identical to those discussed in Section 4.2, remove the derivatives from the dependent variable u by integrating by parts, twice in x and once in t, to arrive at the identity

$$\int_0^\infty \int_0^1 P^{(2)}u(x,t)S_k(x)S_\ell^*(t)w(x)\tau(t)\,dx\,dt$$

$$= \int_0^\infty \int_0^1 u(x,t)\left(-\frac{\partial}{\partial t} - \frac{\partial^2}{\partial x^2}\right)[S_k(x)S_\ell^*(t)w(x)\tau(t)]\,dx\,dt$$

$$+ B_{T_2}$$

where

$$B_{T_2} = \int_0^1 (uS_k S_\ell^* w\tau)(x,t)\Big|_0^\infty dx - \int_0^\infty \left(\frac{\partial u}{\partial x} S_k S_\ell^* w\tau\right)(x,t)\Big|_0^1 dt$$
$$+ \int_0^\infty \left(uS_\ell^*\tau(S_k w)'\right)(x,t)\Big|_0^1 dt.$$

Assuming $B_{T_2} = 0$, replacing u by u_{m_x,m_t}, and following the procedure leading to (5.84) yields the discrete system [6]

$$D(\phi')\left[-\frac{1}{h^2}I_{m_x}^{(2)} - \frac{1}{h}I_{m_x}^{(1)}D\left(\frac{\phi''}{(\phi')^2} + 2\frac{w'}{\phi'w}\right)\right.$$
$$\left. - D\left(\frac{w''}{(\phi')^2 w}\right)\right]D(\phi')V^{(2)} \qquad (6.6)$$
$$+ V^{(2)}D\left(\sqrt{\Upsilon'}\right)\left[-\frac{1}{h}I_{m_t}^{(1)} - D\left(\frac{\tau'}{\tau\Upsilon'}\right)\right]^T D\left(\sqrt{\Upsilon'}\right)$$
$$= G^{(2)}$$

where the coefficients in (6.2) are stored in $(U^{(2)})_{ij} = u_{ij}$ and the point evaluations of the forcing term are in the matrix $(F^{(2)})_{ij} = f(x_i, t_j)$. For the same reasons as in the previous chapter (in particular, for balancing the error contributions), the mesh selections have been made equal in (6.6), i.e., $h = h_x = h_t$. The points (x_j, t_k), where $x_j = \phi^{-1}(jh)$ and $t_k = \Upsilon^{-1}(kh)$, are the inverse images of a uniform grid on the real line under the maps ϕ and Υ, respectively. The matrices $V^{(2)}$ and $G^{(2)}$ are defined by

$$V^{(2)} = D(w)U^{(2)}D\left(\frac{\tau}{\sqrt{\Upsilon'}}\right); \quad G^{(2)} = D(w)F^{(2)}D\left(\frac{\tau}{\sqrt{\Upsilon'}}\right). \qquad (6.7)$$

The heat equation may be viewed as a second-order self-adjoint boundary value problem in space and a first-order initial value problem in time. Then as in (4.60) select $w(x) = 1/\sqrt{\phi'(x)}$, and as in (4.81) select $\tau(t) = \sqrt{\Upsilon'(t)}$. This motivates the choice of weight function [6] for (6.1)

$$w(x)\tau(t) = \sqrt{\frac{\Upsilon'(t)}{\phi'(x)}}. \qquad (6.8)$$

Recalling the definition of the spatial matrix $\mathcal{A}(w)$ in (4.40) and its special form A_x in (5.89) shows that the system in (6.6) may be written

$$A_x V^{(2)} + V^{(2)} B_t^T = G^{(2)} \qquad (6.9)$$

where the temporal matrix is obtained from (4.78) and takes the specific form

$$\mathcal{B}(\sqrt{\Upsilon'}) = -\frac{1}{h}I^{(1)} - D\left(\frac{\tau'}{\tau\Upsilon'}\right) = -\frac{1}{h}I^{(1)} + D\left(\frac{1}{2}\right).$$

The second equality follows from $\tau'/(\tau\Upsilon') = \Upsilon''/(2(\Upsilon')^2) \equiv -1/2$ with Υ given by (6.5). The matrix

$$B_t = D((\Upsilon')^{1/2})\left[-\frac{1}{h}I^{(1)} + D\left(\frac{1}{2}\right)\right]D((\Upsilon')^{1/2}) \qquad (6.10)$$

is just a weighted version of $\mathcal{B}(\sqrt{\Upsilon'})$. The weighted coefficients and forcing term are given by

$$V^{(2)} = D((\phi')^{-1/2})U^{(2)}; \quad G^{(2)} = D((\phi')^{-1/2})F^{(2)}. \qquad (6.11)$$

The following theorem provides the assumptions necessary for this approach to be exponentially accurate. Because of the spatial weight, the stepsize h will be denoted h_s, as in Theorem 4.11.

Theorem 6.1 *Consider the maps* $\phi : D_E \to D_S$ *and* $\Upsilon : D_W \to D_S$ *given in* (6.4) *and* (6.5). *For the weight function*

$$w(x)\tau(t) = \sqrt{\frac{\Upsilon'(t)}{\phi'(x)}}$$

assume that $f/\sqrt{\phi'} \in \mathbf{B}(D_E)$ *and that* $uF \in \mathbf{B}(D_E)$, *where*

$$F = \sqrt{\phi'}\left(1/\phi'\right)'',\ (\phi')^{3/2},\ \phi''/\sqrt{\phi'},\ \left(1/\sqrt{\phi'}\right)''.$$

Also assume that $f\sqrt{\Upsilon'} \in \mathbf{B}(D_W)$ *and that* $uF \in \mathbf{B}(D_W)$, *where*

$$F = \sqrt{\Upsilon'},\ (\Upsilon')^{3/2},\ \Upsilon''/\sqrt{\Upsilon'}.$$

Further assume

$$|u(x,t)| \leq Cx^{\alpha_s+1/2}(1-x)^{\beta_s+1/2}t^{\gamma+1/2}t^{-\delta+1/2}, \qquad (6.12)$$

$$(x,t) \in (0,1)\times(0,\infty).$$

Making the selections

$$N_x = \left[\left|\frac{\alpha_s}{\beta_s}M_x + 1\right|\right],\quad h_s = \left(\frac{\pi d}{\alpha_s M_x}\right)^{1/2} \qquad (6.13)$$

and
$$M_t = \left[\left|\frac{\alpha_s}{\gamma} M_x + 1\right|\right], \quad N_t = \left[\left|\frac{\alpha_s}{\delta} M_x + 1\right|\right] \quad (6.14)$$

results in
$$\|u - u_{m_x,m_t}\|_\infty \le K M_x^2 \exp\left(-(\pi d \alpha_s M_x)^{1/2}\right). \quad (6.15)$$

∎

The formulation (6.9) with A_x and B_t given in (5.89) and (6.10), respectively, may be written in the equivalent form

$$\mathcal{B}^{(2)} \operatorname{co}(V^{(2)}) = \operatorname{co}(G^{(2)}) \quad (6.16)$$

where
$$\mathcal{B}^{(2)} \equiv I_{m_t} \otimes A_x + B_t \otimes I_{m_x} \quad (6.17)$$

and
$$\operatorname{co}(V^{(2)}) = \operatorname{co}\left(D((\phi')^{-1/2}) U^{(2)}\right)$$
$$= \left(I_{m_t} \otimes D((\phi')^{-1/2})\right) \operatorname{co}(U^{(2)})$$

with a similar concatenated equality for $\operatorname{co}(G^{(2)})$.

The equivalent systems (6.9) and (6.16) may be solved in a manner analogous to that discussed in Section 5.4. Indeed, the structure of the system in (6.9) is identical to that of the system in (5.90), if in the latter A_y is replaced by B_t and, at least formally, the Schur decomposition or the full diagonalization procedure described there is applicable to (6.9). Moreover, the alternative derivation beginning with (5.91) is a handy device in the present chapter. There the second boundary value problem (5.92) is replaced by the initial value problem corresponding to the temporal domain in the present case. To the latter, the procedure following (4.77) is applied. All of the matrices (up to a diagonal weighting factor) encountered in this chapter have occurred in Chapter 4 or 5. One difficulty with the previous methods of solution (besides an assumed diagonalizability of B_t) is the necessity of dealing with the complex eigenvalues and eigenvectors of B_t. To circumvent this issue, note that the block structure of $\mathcal{B}^{(2)}$ is the same as the block structure of $\mathcal{A}^{(2)}$ illustrated in Figure 5.2. A scheme designed to take advantage of this structure and to avoid the possible complex arithmetic is a "partial diagonalization" procedure [8]. As in (5.107), let P denote the orthogonal diagonalizer of A_x. Then (via a premultiplication by $I_{m_t} \otimes P^T$) the system (6.16) is equivalent to

$$\mathcal{B}_p^{(2)} \text{co}(W^{(2)}) = \text{co}(H^{(2)}) \tag{6.18}$$

where $\mathcal{B}_p^{(2)}$ is the partially diagonalized $\mathcal{B}^{(2)}$ defined by

$$\mathcal{B}_p^{(2)} = I_{m_t} \otimes \Lambda_x + B_t \otimes I_{m_x} \tag{6.19}$$

and Λ_x is the diagonal matrix of eigenvalues of A_x. The concatenated matrix of coefficients and forcing vectors are, respectively,

$$\text{co}(W^{(2)}) = (I_{m_t} \otimes P^T)\text{co}(V^{(2)})$$

and

$$\text{co}(H^{(2)}) = (I_{m_t} \otimes P^T)\text{co}(G^{(2)}).$$

The $m_x m_t \times m_x m_t$ matrix $\mathcal{B}_p^{(2)}$ can be pictorially displayed as

$$\begin{bmatrix} \Lambda_x + b_{-M_t,-M_t} I_{m_x} & \cdots & b_{-M_t,N_t} I_{m_x} \\ b_{-M_t+1,-M_t} I_{m_x} & & \vdots \\ \vdots & \ddots & \vdots \\ b_{N_t,-M_t} I_{m_x} & \cdots & \Lambda_x + b_{N_t,N_t} I_{m_x} \end{bmatrix}$$

where the m_t^2 numbers (b_{ij}), $-M_t \leq i,j \leq N_t$ are the entries in the matrix B_t. That is, each block of $\mathcal{B}_p^{(2)}$ is a diagonal matrix. This structure requires that only $m_x(m_t^2 - 2m_t + 1)$ nonzero elements be stored. A method of solution of (6.18) is a block Gauss–Jordan routine. At one level of programming the blocks of $\mathcal{B}_p^{(2)}$ are treated as if they were constants so that divisions are simply inverses of diagonal matrices and the system (6.18) is naturally suited for vector computation [8].

Before turning to the implementation of the above procedure on an example, there is a method to reduce the size of any of the above systems in a number of frequently occurring cases. The assumed decay rate in (6.12) forces the selection of the number of temporal nodes in (6.14). This selection is mandatory to achieve the convergence rate in (6.15) if the solution of the partial differential equation has rational behavior in the temporal domain. However, many of the problems considered in this chapter exhibit an exponential rate of decrease, and for these problems the selection in (6.14) is unnecessarily large. At this point, there is a distinction between the maps $\ell n(t)$ and $\ell n(\sinh(t))$. In Exercise 6.1 the exponential convergence rate in (6.15) is attained for the former but not for the latter. As in the case of ordinary differential equations (Example 4.13), there is an economization in the size of the discrete system if the solution decays exponentially. The following example illustrates this situation. The analogy continues in the sense that one could obtain the results in this example using the map $\ell n(\sinh(t))$ but the system sizes are much larger.

Example 6.2 To examine this economization define the system

$$\begin{aligned}\frac{\partial u}{\partial t}(x,t) - \frac{\partial^2 u}{\partial x^2}(x,t) &= [x(1-x)(1-t)+2t]e^{-t}\\ u(0,t) &= u(1,t) = 0, \quad t>0 \\ u(x,0) &= 0, \quad 0<x<1\end{aligned} \quad (6.20)$$

which has true solution

$$u(x,t) = x(1-x)te^{-t}.$$

The assumed form of the bound for this exponentially decaying case reads

$$|u(x,t)| \le C x^{\alpha_s+1/2}(1-x)^{\beta_s+1/2}t^{\gamma+1/2}e^{-\delta t}, \qquad (6.21)$$

instead of (6.12). From (4.74) the choice for N_t in (6.14) is replaced by

$$N_e = \left[\left|\frac{1}{h_s}\ell n\left(\frac{\alpha_s}{\delta}M_x h_s\right)+1\right|\right]. \qquad (6.22)$$

Since $\alpha_s = \beta_s = \gamma = 1/2$ and $\delta = 1$ (6.13) yields $h_s = \pi/\sqrt{M_x}$ $(d = \pi/2)$ and $M_x = N_x$. Further, (6.14) gives $M_x = M_t$ and

$$N_e = \left[\left|\frac{\sqrt{M_x}}{\pi}\ell n\left(\frac{\pi\sqrt{M_x}}{2}\right) + 1\right|\right]$$

follows from (6.22). The discrete system for this problem is given by (6.9), namely,

$$A_x V^{(2)} + V^{(2)} B_t^T = G^{(2)}$$

where

$$A_x = D\left(\frac{1}{x(1-x)}\right)\left[-\frac{1}{h_s^2}I^{(2)} + D\left(\frac{1}{4}\right)\right]D\left(\frac{1}{x(1-x)}\right)$$

$$B_t = D\left(\frac{1}{\sqrt{t}}\right)\left[-\frac{1}{h_s}I^{(1)} + D\left(\frac{1}{2}\right)\right]^T D\left(\frac{1}{\sqrt{t}}\right)$$

and

$$V^{(2)} = D\left(\sqrt{x(1-x)}\right)U^{(2)}; \quad G^{(2)} = D\left(\sqrt{x(1-x)}\right)F^{(2)}.$$

The solution $U^{(2)}$ gives the coefficients in the approximate Sinc-Galerkin solution

$$u_{m_x, m_t}(x, t) = \sum_{j=-M_t}^{N_e} \sum_{i=-M_x}^{N_x} u_{ij} S_{ij}(x, t)$$

where

$$S_{ij}(x, t) = S(i, h_s) \circ \left(\ell n\left(\frac{x}{1-x}\right)\right) S(j, h_s) \circ (\ell n(t)).$$

In Table 6.1 the notation $\|E_S(h_s)\|$ is the error on the sinc grid. The column headed $\mathcal{B}^{(2)}$ is the size of the matrix in (6.17) used to compute the solution. If N_t is selected according to (6.14), as opposed to the selection in (6.22), the numbers in this column are replaced by $9^2 = 81$, $17^2 = 289$, $33^2 = 1089$, and $65^2 = 4225$. For example, this would be the size of the system if the map $\Upsilon(t) = \ell n(\sinh(t))$ had been used instead of $\ell n(t)$. The final time level where the error $\|E_S(h_s)\|$ is computed $(t_{N_e} = \exp(N_e h_s) = 9.21$ in the last row $(N_e = 4))$ would be replaced by $t_{N_t} = \exp(N_t h_s) \approx 5 \times 10^6$, corresponding to the choice $N_t = 32$. The method works just fine with this choice, but it is a bit of a waste. The selection (6.22) defines a cut off which corresponds to the limits of the accuracy of the method. Thus there is little point in adding the extra temporal nodes to the approximation. No matter how the final temporal node is selected, the approximate solution u_{m_x,m_t} is globally defined and, should it be desired, can be evaluated at any positive time.

$M_x = N_x$	M_t	N_e	$\mathcal{B}^{(2)}$	$h_s = \pi/\sqrt{M_x}$	$\|E_S(h_s)\|$
4	4	1	45	1.571	$.925 \times 10^{-3}$
8	8	2	187	1.111	$.273 \times 10^{-3}$
16	16	3	660	0.785	$.842 \times 10^{-5}$
32	32	4	2368	0.555	$.704 \times 10^{-6}$

Table 6.1 Errors and system sizes for the one-dimensional heat equation in (6.20).

∎

The development for the heat equation in two space dimensions combines the preceding temporal development with the discussion of the two-dimensional Poisson problem in Section 5.4. Both the formulation as a Sylvester equation or an equivalent Kronecker sum will be given and a companion solution technique outlined. The development here is found in [12].

The two-dimensional form of (6.1) reads

$$P^{(3)}u(x,y,t) \equiv \frac{\partial u(x,y,t)}{\partial t} - \Delta^{(2)} u(x,y,t) = f(x,y,t)$$
$$u(x,y,0) = 0, \quad (x,y) \in S = (0,1) \times (0,1) \quad (6.23)$$
$$u(x,y,t)|_{\partial S} = 0, \quad t > 0.$$

The approximate solution is given by

$$u_{m_x,m_y,m_t}(x,y,t) = \sum_{k=-M_t}^{N_t} \sum_{j=-M_y}^{N_y} \sum_{i=-M_x}^{N_x} u_{ijk} S_{ijk}(x,y,t)$$

where

$$S_{ijk}(x,y,t) \equiv S_i(x)S_j(y)S_k^*(t)$$
$$\equiv S(i,h) \circ \phi(x) S(j,h) \circ \phi(x) S(k,h) \circ \Upsilon(t)$$

and $m_q = M_q + N_q + 1$ for $q = x$, y or t. The maps for the basis functions are the same as in (6.4) and (6.5) for $S_i(x)$ and $S_k^*(t)$, respectively. The function $S_j(y)$ is simply a copy of $S_i(x)$ in the y direction. The inner product is defined to be

$$(f,g) = \int_0^\infty \int_0^1 \int_0^1 f(x,y,t) g(x,y,t) \left(\frac{\Upsilon'(t)}{\phi'(x)\phi'(y)}\right)^{1/2} dx\, dy\, dt$$

where the weight function is the three-dimensional version of (6.8).

Integration by parts is again used to remove the derivatives from u appearing in the orthogonalization of the residual

$$\left(P^{(3)}u - f, S_{pqr}\right) = 0.$$

After replacing u by u_{m_x, m_y, m_t}, this leads to the following form for the discrete system

$$\mathcal{A}^{(2)} \text{mat}\left(V^{(3)}\right) + \text{mat}\left(V^{(3)}\right) B_t^T = \text{mat}\left(G^{(3)}\right) \qquad (6.24)$$

where $\mathcal{A}^{(2)}$ is defined in (5.109), and

$$\text{mat}\left(V^{(3)}\right) = \left[D\left((\phi_y')^{-1/2}\right) \otimes D\left((\phi_x')^{-1/2}\right)\right] \text{mat}\left(U^{(3)}\right)$$

and

$$\text{mat}\left(G^{(3)}\right) = \left[D\left((\phi_y')^{-1/2}\right) \otimes D\left((\phi_x')^{-1/2}\right)\right] \text{mat}\left(F^{(3)}\right).$$

The matrix operator "mat" is found in Definition A.30. The system (6.24) is the same as (5.117) if A_z is replaced by B_t^T. A theorem analogous to Theorem 6.1 holds with the assumptions there stated for the x variable assumed to also hold for the y variable.

The formulation (6.24) has the equivalent representation (see (5.125) and therein replace A_z by B_t):

$$\mathcal{B}^{(3)} \text{co}\left(V^{(3)}\right) = \text{co}\left(G^{(3)}\right) \qquad (6.25)$$

where
$$\mathcal{B}^{(3)} = I_{m_t} \otimes \mathcal{A}^{(2)} + B_t \otimes I_{m_y m_x}. \qquad (6.26)$$

The vectors $\mathrm{co}(V^{(3)})$ and $\mathrm{co}(G^{(3)})$ are given by

$$\begin{aligned}\mathrm{co}\left(V^{(3)}\right) &= \left[I_{m_t} \otimes D\left((\phi'_y)^{-1/2}\right) \otimes D\left((\phi'_x)^{-1/2}\right)\right] \mathrm{co}\left(U^{(3)}\right) \\ &= (I_{m_t} \otimes D_{yx})\,\mathrm{co}\left(U^{(3)}\right)\end{aligned}$$

and
$$\mathrm{co}\left(G^{(3)}\right) = (I_{m_t} \otimes D_{yx})\,\mathrm{co}\left(F^{(3)}\right).$$

If P and Q denote the orthogonal diagonalizers of A_x and A_y, respectively, then the system (6.25) is also equivalent to

$$\mathcal{B}_p^{(3)}\mathrm{co}(W^{(3)}) = \mathrm{co}(H^{(3)}) \qquad (6.27)$$

where $\mathcal{B}_p^{(3)}$ is the partially diagonalized $\mathcal{B}^{(3)}$ defined by

$$\mathcal{B}_p^{(3)} = I_{m_t} \otimes \Lambda_{yx} + B_t \otimes I_{m_y m_x}$$

and
$$\mathrm{co}(W^{(3)}) = \left(I_{m_t} \otimes Q^T \otimes P^T\right) \mathrm{co}\left(V^{(3)}\right)$$

and
$$\mathrm{co}(H^{(3)}) = \left(I_{m_t} \otimes Q^T \otimes P^T\right) \mathrm{co}\left(G^{(3)}\right).$$

The matrix Λ_{yx} is the Kronecker sum (Definition A.26) of the matrices Λ_y and Λ_x of eigenvalues of A_y and A_x, respectively. The $m_x m_y m_t \times m_x m_y m_t$ matrix $\mathcal{B}_p^{(3)}$ has a structure that is identical to the structure of the $m_x m_t \times m_x m_t$ matrix $\mathcal{B}_p^{(2)}$ in (6.19). Indeed, $\mathcal{B}_p^{(3)}$ can be written as

$$\begin{bmatrix} \Lambda_{yx} + b_{-M_t,-M_t}I_{m_x m_y} & \cdots & b_{-M_t,N_t}I_{m_x m_y} \\ b_{-M_t+1,-M_t}I_{m_x m_y} & & \vdots \\ \vdots & \ddots & \vdots \\ b_{N_t,-M_t}I_{m_x m_y} & \cdots & \Lambda_{yx} + b_{N_t,N_t}I_{m_x m_y} \end{bmatrix}$$

where the m_t^2 numbers (b_{ij}) are the entries in B_t. The display shows that the matrix $B_p^{(3)}$ has only $m_x m_y (m_t^2 - 2m_t + 1)$ nonzero elements to be stored. The off-diagonal blocks of $B_p^{(3)}$ are all scalar diagonal matrices. Again, as was the case for $B_p^{(2)}$, a block Gauss–Jordan technique with vector architecture is a viable and efficient method to take advantage of this structure.

Example 6.3 The solution to the problem

$$\begin{aligned}\frac{\partial u(x,y,t)}{\partial t} - \Delta^{(2)} u(x,y,t) &= f(x,y,t) \\ u(x,y,0) &= 0, \quad (x,y) \in S = (0,1) \times (0,1) \\ u(x,y,t)|_{\partial S} &= 0, \quad t > 0\end{aligned} \qquad (6.28)$$

exhibits a logarithmic singularity at the origin and decays exponentially in time so that it is sort of a product of the problems (one-dimensional) worked out in Examples 4.12 and 4.14. The forcing term $f(x,y,t)$ is selected so that the solution is given by

$$u(x,y,t) = x \ln(x) y \ln(y) t^{3/2} e^{-t}.$$

The extension of (6.21) for this example reads (since there is exponential decay in time)

$$\begin{aligned}|u(x,y,t)| \leq {} & C x^{\alpha_s + 1/2} (1-x)^{\beta_s + 1/2} y^{\tau_s + 1/2} \\ & (1-y)^{\eta_s + 1/2} t^{\gamma + 1/2} e^{-\delta t}.\end{aligned}$$

Here $\alpha_s = \beta_s = \tau_s = \eta_s = 1/2$ and $\gamma = \delta = 1$. Thus $h_s = \pi/\sqrt{M_x}$ ($d = \pi/2$) and $M_x = N_x = M_y = N_y$, $M_t = M_x/2$ and, using (6.22),

$$N_e = \left[\left|\frac{\sqrt{M_x}}{\pi} \ln\left(\frac{\pi \sqrt{M_x}}{2}\right) + 1\right|\right].$$

The results listed in Table 6.2 were computed using the partially diagonalized system in (6.27).

M_x	N_x	M_y	N_y	M_t	N_e	$h_s = \pi/\sqrt{M_x}$	$\|E_S(h_s)\|$
4	4	4	4	2	1	1.571	$.968 \times 10^{-3}$
8	8	8	8	4	2	1.111	$.303 \times 10^{-3}$
16	16	16	16	8	3	0.785	$.394 \times 10^{-4}$

Table 6.2 Errors in the two-dimensional heat problem in (6.28).

∎

The method of solution for (6.9) (equivalently (6.16)) and their higher dimensional analogues (6.24) (equivalently (6.25)) has been purely descriptive. The invertibility of $\mathcal{B}^{(2)}$ in (6.16) is equivalent to the unique solvability of (6.9). This unique solvability is most accessible from the Sylvester formulation. To ensure invertibility of $\mathcal{B}^{(2)}$ it suffices, by equation (6.17) and Theorem A.27, to verify that the matrix B_t has no negative real spectrum. This looks reasonable as a part of the summand defining B_t is skew-symmetric ($I^{(1)}$ is, by Theorem 4.19, skew-symmetric) and, by Theorem 5.8, A_x has positive real spectrum. Since A_y also has positive real spectrum, then the same argument (with Theorem A.23) will imply that the system (6.25), with the coefficient matrix $\mathcal{B}^{(3)}$, is uniquely solvable.

To see that the spectrums are disjoint, write B_t in (6.10) in the form
$$B_t = C_1 + C_2$$
where
$$C_1 = -D\left(\sqrt{\Upsilon'}\right)\left(\frac{1}{h}I^{(1)}\right)D\left(\sqrt{\Upsilon'}\right)$$
is skew-symmetric (the diagonal matrices are real) and
$$C_2 = \frac{1}{2}D\left(\sqrt{\Upsilon'}\right)D\left(\frac{-\Upsilon''}{(\Upsilon')^2}\right)D\left(\sqrt{\Upsilon'}\right) = \frac{1}{2}D\left(\frac{1}{t}\right)$$
is symmetric. If $(\lambda_t, \vec{v})$ is an eigenpair of B_t, then, upon adding the identities
$$\vec{v}^* B_t \vec{v} = \lambda_t \vec{v}^* \vec{v}$$

and
$$\vec{v}^* B_t^T \vec{v} = \bar{\lambda}_t \vec{v}^* \vec{v}.$$
it follows that
$$\vec{v}^* (C_2) \vec{v} = \frac{1}{2}\vec{v}^* \left(B_t + B_t^T\right) \vec{v} = Re(\lambda_t)\vec{v}^* \vec{v}.$$

Since the diagonal entries in C_2 are positive, it follows that $Re(\lambda_t)$ is positive. Since B_t has spectrum with positive real part and the spectrum of A_x is positive (real), it follows that $\mathcal{B}^{(2)} = I \otimes A_x + B_t \otimes I$ has spectrum with positive real part and is invertible. The only change in the above argument for the map $\Upsilon(t) = \ell n(\sinh(t))$ is that the diagonal matrix C_2 is defined by a different function (Exercise 6.2).

Finally, an upper bound for points in the spectrum of $\mathcal{B}^{(2)}$ is obtained from the preceding argument and Theorem 5.8. If $\lambda \in \sigma\left(\mathcal{B}^{(2)}\right)$, then there are eigenvalues λ_x and λ_t of A_x and B_t, respectively, so that $\lambda = \lambda_x + \lambda_t$. Hence,

$$\begin{aligned} |\lambda| \geq Re(\lambda) &= Re(\lambda_x + \lambda_t) \\ &= \lambda_x + Re(\lambda_t) \\ &\geq \frac{4}{(b-a)^2} \\ &= 4. \end{aligned}$$

The last inequality is from Theorem 5.8, and the last equality is due to the present convention, which has $a = 0$ and $b = 1$. By inversion it follows that
$$\frac{1}{|\lambda|} \leq \frac{(b-a)^2}{4} = \frac{1}{4}. \tag{6.29}$$

This inequality bounds the spectral radius of the inverse of $\mathcal{B}^{(2)}$ and will play a suggestive role in Section 6.3.

6.2 Treatment of Boundary and Initial Conditions

The treatment of problems with nonhomogeneous Dirichlet boundary conditions and nonhomogeneous initial conditions is completely

TIME-DEPENDENT PROBLEMS

analogous to the methodology described in Section 4.4. To keep the analogy as close as possible, the spatial interval will be designated as (a, b). To illustrate the situation, consider the problem

$$P^{(2)}u(x,t) \equiv \frac{\partial u}{\partial t}(x,t) - \frac{\partial^2 u}{\partial x^2}(x,t) = f(x,t)$$
$$u(a,t) = \gamma(t), \quad u(b,t) = \delta(t), \quad t \geq 0 \qquad (6.30)$$
$$u(x,0) = g(x), \quad a \leq x \leq b.$$

where $\gamma(0) = g(a)$ and $\delta(0) = g(b)$. The approximate solution is written as

$$u_{m_x,m_t}(x,t) = \sum_{j=-M_t-1}^{N_t} \sum_{i=-M_x-1}^{N_x+1} u_{ij}\chi_i(x)\theta_j(t)$$

where two linear functions are added to the sinc basis in the spatial dimension

$$\chi_i(x) = \begin{cases} \dfrac{b-x}{b-a}, & i = -M_x - 1 \\ S(i,h) \circ \phi(x), & -M_x \leq i \leq N_x \\ \dfrac{x-a}{b-a}, & i = N_x + 1 \end{cases} \qquad (6.31)$$

and one rational function is appended to the temporal base

$$\theta_j(t) = \begin{cases} \dfrac{t+1}{t^2+1}, & j = -M_t - 1 \\ S(j,h) \circ \Upsilon(t), & -M_t \leq j \leq N_t. \end{cases} \qquad (6.32)$$

As in the last section, the mesh sizes h_x and h_t have been set equal and denoted by h. Interpolating the boundary and initial conditions in (6.30) dictates that

$$u_{m_x,m_t}(a,t) = \sum_{j=-M_t-1}^{N_t} u_{-M_x-1,j}\theta_j(t) = \gamma(t)$$

$$u_{m_x,m_t}(b,t) = \sum_{j=-M_t-1}^{N_t} u_{N_x+1,j}\theta_j(t) = \delta(t)$$

$$u_{m_x,m_t}(x,0) = \sum_{i=-M_x-1}^{N_x+1} u_{i,-M_t-1}\chi_i(x) = g(x).$$

The sinc approximation to (6.30) is defined by

$$u_{m_x,m_t}(x,t) = \sum_{j=-M_t}^{N_t} \sum_{i=-M_x}^{N_x} u_{ij} S_{ij}(x,t) + \bar{\gamma}(t)\chi_{-M_x-1}(x) \\ + \bar{\delta}(t)\chi_{N_x+1}(x) + g(x)\theta_{-M_t-1}(t) \quad (6.33)$$

where

$$\bar{\gamma}(t) = \gamma(t) - g(a)\theta_{-M_t-1}(t) \\ \bar{\delta}(t) = \delta(t) - g(b)\theta_{-M_t-1}(t).$$

Since $\bar{\gamma}(t)$ and $\bar{\delta}(t)$ are known functions, the orthogonalization of the residual

$$\left(P^{(2)}u_{m_x,m_t} - f, S_{k\ell}\right) = 0$$

for $-M_x \leq k \leq N_x, -M_t \leq \ell \leq N_t$ may be written

$$\left(P^{(2)}u_h - \bar{f}, S_{k\ell}\right) = 0 \quad (6.34)$$

where the homogeneous part of the approximate solution is given by

$$u_h(x,t) = \sum_{j=-M_t}^{N_t} \sum_{i=-M_x}^{N_x} u_{ij} S_{ij}(x,t). \quad (6.35)$$

The $S_{ij}(x,t)$ are defined in (6.3), and $\bar{f}$ is given by

$$\bar{f}(x,t) = f(x,t) - P^{(2)}\left[\bar{\gamma}(t)\chi_{-M_x-1}(x) \\ + \bar{\delta}(t)\chi_{N_x+1}(x) + g(x)\theta_{-M_t-1}(t)\right]. \quad (6.36)$$

Choosing the weight function in (6.8) leads to the discrete system

$$A_x V^{(2)} + V^{(2)} B_t^T = G^{(2)} \quad (6.37)$$

where $V^{(2)}$ and $G^{(2)}$ are again given by (6.11). In the latter the matrix $F^{(2)}$ is now the matrix of point evaluations of $\bar{f}$ in (6.36).

TIME-DEPENDENT PROBLEMS

In the case of homogeneous initial conditions ($g \equiv 0$), $\bar{f}$ in (6.36) is replaced by

$$\hat{f}(x,t) = f(x,t) - P^{(2)}\left[\gamma(t)\frac{b-x}{b-a} + \delta(t)\frac{x-a}{b-a}\right]. \qquad (6.38)$$

The system for this problem is the same as the system in (6.37), where the matrix $G^{(2)}$ is again given by (6.11) and $F^{(2)}$ is now the matrix of point evaluations of $\hat{f}$. To reinforce the ongoing analogy with ordinary differential equations, compare (6.38) with the transformation in Exercise 4.7. The function $\hat{f}$ is playing the same role here that it played in that nonhomogeneous Dirichlet problem.

The most expedient method of dealing with radiation boundary conditions is, as discussed in the introduction of Section 4.4, done by assuming homogeneous initial conditions ($g(x) \equiv 0$ in (6.30)). There is no loss of generality due to Exercise 6.3. The problem dealt with here is defined by the system

$$\begin{aligned} P^{(2)}u(x,t) &\equiv \frac{\partial u}{\partial t}(x,t) - \frac{\partial^2 u}{\partial x^2}(x,t) = f(x,t) \\ a_1 u_x(a,t) &- a_0 u(a,t) = 0, \quad t \geq 0 \\ b_1 u_x(b,t) &+ b_0 u(b,t) = 0, \quad t \geq 0 \\ u(x,0) &= 0, \quad a \leq x \leq b. \end{aligned} \qquad (6.39)$$

Following the development in Section 4.4, the approximate solution is defined by

$$u_A(x,t) = \sum_{j=-M_x-1}^{N_x+1} \sum_{k=-M_t}^{N_t+1} c_{jk}\xi_j(x)\zeta_k(t) \qquad (6.40)$$

where

$$\xi_j(x) \equiv \begin{cases} \omega_a(x), & j = -M_x - 1 \\ \frac{S(j,h) \circ \phi(x)}{\phi'(x)}, & j = -M_x, \ldots, N_x \\ \omega_b(x), & j = N_x + 1 \end{cases} \qquad (6.41)$$

and

$$\zeta_k(t) \equiv \begin{cases} S(k,h) \circ \Upsilon(t), & k = -M_t, \ldots, N_t \\ \omega_\infty(t), & k = N_t + 1 \end{cases}. \qquad (6.42)$$

The boundary basis functions $\omega_a, \omega_b,$ and ω_∞ are

$$\omega_a(x) = a_0 \frac{(x-a)(b-x)^2}{(b-a)^2} + a_1 \frac{(2x+b-3a)(b-x)^2}{(b-a)^3}$$

$$\omega_b(x) = b_1 \frac{(-2x+3b-a)(x-a)^2}{(b-a)^3} + b_0 \frac{(b-x)(x-a)^2}{(b-a)^2}$$

and

$$\omega_\infty(t) \equiv \frac{t}{t+1}.$$

To determine the discrete system, the residual is orthogonalized against the set $\{\psi_p(x)\theta_q(t)\}$. Referring to Section 4.4, these functions are chosen to be

$$\psi_p(x) \equiv \frac{S(p,h) \circ \phi(x)}{\phi'(x)}, \quad p = -M_x - 1, \ldots, N_x + 1 \qquad (6.43)$$

and

$$\theta_q(t) \equiv S(q,h) \circ \Upsilon(t), \quad q = -M_t, \ldots, N_t + 1. \qquad (6.44)$$

There is a slight overlap in this notation and that in (6.32), but for the indices that are the same the functions are the same. The choice $g(x) \equiv w(x) \equiv 1$ is made so that the $\xi_j(x)$ in (6.41) are the same as those used in (4.94). It is convenient to first derive the discrete system for the heat equation with homogeneous Dirichlet boundary conditions. In this setting the additional basis functions ω_l, ($l = a, b, \infty$) are not needed in (6.40). Hence the assumed approximate solution to (6.39) with Dirichlet boundary conditions ($a_1 = b_1 = 0$) takes the form

$$u_{m_x, m_t}(x, t) = \sum_{j=-M_x}^{N_x} \sum_{k=-M_t}^{N_t} c_{jk} \xi_j(x) \zeta_k(t)$$

$$(6.45)$$

$$= \sum_{j=-M_x}^{N_x} \sum_{k=-M_t}^{N_t} c_{jk} \frac{S(j,h) \circ \phi(x)}{\phi'(x)} S(k,h) \circ \Upsilon(t).$$

Integrating by parts twice in x and once in t to remove the derivatives from u and then substituting (6.45) in for u leads, via a lengthy calculation, to the matrix equation

$$A_w CD\left(\frac{1}{\sqrt{\Upsilon'}}\right) + D\left(\frac{1}{(\phi')^3}\right)CB_w^T = F_w. \qquad (6.46)$$

Here the $m_x \times m_x$ matrix A_w is a weighted multiple of $\mathcal{A}(1/\phi')$ in (4.44) given by

$$A_w = \mathcal{A}\left(\frac{1}{\phi'}\right)D\left(\frac{1}{\phi'}\right) \qquad (6.47)$$

while the $m_t \times m_t$ matrix B_w is from (6.10)

$$B_w = D\left(\frac{1}{\sqrt{\Upsilon'}}\right)B_t = \left[\frac{-1}{h}I^{(1)} + D\left(\frac{1}{2}\right)\right]D(\sqrt{\Upsilon'}) \qquad (6.48)$$

and the $m_x \times m_t$ matrix F_w is

$$F_w = D\left(\frac{1}{(\phi')^2}\right)F^{(2)}D\left(\frac{1}{\sqrt{\Upsilon'}}\right). \qquad (6.49)$$

The jk-th entry of the $m_x \times m_t$ matrix $F^{(2)}$, for $j = -M_x, \ldots, N_x$ and $k = -M_t, \ldots, N_t$ contains the point evaluations of the function $f(x,t)$. The coefficients $\{c_{jk}\}$ are the jk-th entries of the $m_x \times m_t$ matrix C. Up to the additional factor of $1/\phi'$ in the definition of the basis (6.41), this system is the one in (6.6) where $w = 1/\phi'$ and $\tau = \sqrt{\Upsilon'}$.

When (6.45) is replaced by the approximate solution in (6.40) for the full radiation boundary conditions in (6.39), the extra inner products contributing to the above matrix formulation (6.46) consist of point evaluations of $w_l, (l = a, b, \infty)$ and their derivatives. This is because the orthogonalization of the residual is done with respect to the set $\{\psi_p(x)\theta_q(t)\}$ defined in (6.43) and (6.44). The system that results from this approach is given by

$$\mathcal{A}C\mathcal{D}_t + \mathcal{D}_s C B^T = \mathcal{F}. \qquad (6.50)$$

Here the $(m_x + 2) \times (m_t + 1)$ matrix $\mathcal{F}$ has the same form as F_w in (6.49), where $F^{(2)}$ is replaced by the $(m_x+2)\times(m_t+1)$ matrix consisting of the point evaluations of the function $f(x,t)$ at the sinc meshpoints (x_j, t_k) for $j = -M_x - 1, \ldots, N_x + 1$ and $k = -M_t, \ldots, N_t + 1$.

The bordered matrices $\mathcal{A}, \mathcal{B}, \mathcal{D}_s$, and $\mathcal{D}_t$ have the following forms. The $(m_x+2) \times (m_x+2)$ matrices $\mathcal{A}$ (see (4.98)) and $\mathcal{D}_s$ are given by

$$\mathcal{A} \equiv \left[\vec{a}_{-M_x-1} \mid \tilde{A}_w \mid \vec{a}_{N_x+1} \right] \qquad (6.51)$$

$$\mathcal{D}_s \equiv \left[\vec{d}_{-M_x-1} \mid \mathcal{D}\left(\frac{1}{(\phi')^3}\right) \mid \vec{d}_{N_x+1} \right] \qquad (6.52)$$

while the $(m_t+1) \times (m_t+1)$ matrices $\mathcal{B}$ (see (4.89)) and $\mathcal{D}_t$ are given by

$$\mathcal{B} \equiv \left[\tilde{B}_w \mid \vec{b}_{N_t+1} \right] \qquad (6.53)$$

and

$$\mathcal{D}_t \equiv \left[\begin{array}{c|c} \mathcal{D}\left(\frac{1}{\sqrt{\Upsilon'}}\right) & \vec{0} \\ \hline \vec{d}_{N_t+1}^T & \end{array} \right]. \qquad (6.54)$$

The matrix $\tilde{A}_w$ is an $(m_x+2) \times m_x$ copy of A_w in (6.47) and $\tilde{B}_w$ is an $(m_t+1) \times m_t$ copy of B_w in (6.48). The completion of the system in (6.50) is obtained upon defining the j-th component ($j = -M_x-1, \ldots, N_x+1$) or the k-th component ($k = -M_t, \ldots, N_t+1$) of the vectors forming columns or rows of the above matrices by

$$(\vec{a}_{-M_x-1})_j = \left(\frac{\omega_a''}{(\phi')^2}\right)(x_j)$$

$$(\vec{a}_{N_x+1})_j = \left(\frac{\omega_b''}{(\phi')^2}\right)(x_j)$$

$$(\vec{d}_{-M_x-1})_j = \left(\frac{\omega_a}{(\phi')^2}\right)(x_j)$$

$$(\vec{d}_{N_x+1})_j = \left(\frac{\omega_b}{(\phi')^2}\right)(x_j)$$

$$(\vec{b}_{N_t+1})_k = \left(\frac{-\omega_\infty'}{\sqrt{\Upsilon'}}\right)(t_k)$$

and

$$(\vec{d}_{N_t+1})_k = \left(\frac{\omega_\infty}{\sqrt{\Upsilon'}}\right)(t_k).$$

It may help to note that if $\mathcal{B}$ in (6.50) is set equal to zero, $\mathcal{D}_t = I$, and $\mathcal{C}$ and $\mathcal{F}$ are thought of as $(m_x + 2) \times 1$ vectors, then the system in (6.50) is the one-dimensional system in (4.97).

One method of solution for the system in (6.50) is to use the generalized Schur algorithm [3] on the pairs $(\mathcal{A}, \mathcal{D}_s)$ and $(\mathcal{D}_t, \mathcal{B})$, which reduces (6.50) to a form where recursive solution of upper triangular systems is all that is required. Although a diagonalization procedure as outlined in Section 5.4 is not the recommended approach for the solution of the system in (6.50), it is easy to implement and is the method that is used to obtain the results in the next example.

Example 6.4 The following heat equation with radiation boundary conditions will now be considered

$$\begin{aligned}
u_t(x,t) &= u_{xx}(x,t) + f(x,t) \\
2u_x(0,t) - 3u(0,t) &= 0, \quad t \geq 0 \\
u_x(1,t) + 2u(1,t) &= 0, \quad t \geq 0 \\
u(x,0) &= 0, \quad 0 \leq x \leq 1.
\end{aligned} \quad (6.55)$$

If the forcing term $f(x,t)$ is

$$f(x,t) = [(1 + \frac{1}{2}x - x^2) + t(1 - \frac{7}{2}x - x^2)]e^{x-t}$$

then the solution of (6.55) is given by

$$u(x,t) = [1 + \frac{1}{2}x - x^2]te^{x-t}.$$

The coefficients $\{c_{jk}\}$ for the assumed approximate solution in (6.40) are computed from (6.50). The basis element $\omega_\infty(t)$ could be removed from the system if one assumes that the true solution goes to zero as t goes to infinity. This amounts to deleting the final column in (6.53) and the last row in (6.54). The situation here is the same as it was in the discussion following Table 4.11. There is little point in adding extra (unneeded) basis elements to the system. The results in Table 6.3 are the same whether or not ω_∞ is included in the expansion basis. The bound in (6.21) for the case of the spatial weight $w(x) = 1/\phi'(x)$ takes the form

$$|u(x,t)| \leq Cx^\alpha(1-x)^\beta t^{\gamma+1/2}e^{-\delta t}. \quad (6.56)$$

Since $\alpha = \beta = \gamma = 1$ and $\delta = 1$, (6.13) yields $h = \pi/\sqrt{2M_x}$ ($d = \pi/2$), $M_x = N_x = M_t$, and the economized choice N_e is given in (6.22). In each of the mentioned formulas, the parameters with a subscripted s are replaced by their counterparts, defined at the opening of this paragraph.

M_x	N_x	M_t	N_e	$h = \pi/\sqrt{2M_x}$	$\|E_S(h)\|$
4	4	4	1	1.111	$.903 \times 10^{-2}$
8	8	8	2	0.785	$.362 \times 10^{-2}$
16	16	16	3	0.555	$.780 \times 10^{-3}$

Table 6.3 The solution of the heat equation with full radiation boundary conditions in (6.55).

■

In the preceding example the computational scheme is independent of the introduction of an extra boundary basis function (ω_∞ in Example 6.4). For the calculation required in the next example, ω_∞ must be included in the assumed approximate solution (6.40) [7].

Example 6.5 This example illustrates the ability of the assumed approximate solution (6.40) to follow the family of solutions defined by the problem (6.39) when a boundary condition is allowed to change. Specifically, if the selections $f(x,t) \equiv 0, a_1 = b_1 = 1, a_0 = \rho$, and $b_0 = 0$ are made in (6.39) and the initial data is defined by $g(x) = (1 - \cos(x))/2$, then the system takes the form

$$\begin{aligned} u_t(x,t) &= u_{xx}(x,t) \\ u_x(0,t) &= \rho u(0,t), \quad t \geq 0 \\ u_x(1,t) &= 0, \quad t \geq 0 \\ u(x,0) &= \frac{1}{2}(1 - \cos(\pi x)), \quad 0 \leq x \leq 1. \end{aligned} \qquad (6.57)$$

TIME-DEPENDENT PROBLEMS 251

The problem is first transformed to one with homogeneous initial conditions using the transformation in Exercise 6.3. For this transformed problem the coefficients $\{c_{jk}\}$ for the assumed approximate solution in (6.40) are computed from (6.50) for $M \equiv M_x = N_x = N_t = M_t = 16$. In keeping with the remarks made in Example 4.16, the parameters α, β, γ, and δ are taken equal to one. The mesh size is $h = \pi/\sqrt{2M}$. The only alteration in the discrete system (6.50) when ρ is changed is in the bordering vectors $\vec{a}_{-M-1}$ and $\vec{d}_{-M-1}$ in the matrices $\mathcal{A}$ in (6.51) and $\mathcal{D}_s$ in (6.52), respectively. In other words, only the vectors involving $\omega_0(x)$ are changed.

Corresponding to different values of the positive parameter ρ, the solution of the system (6.57), $u_\rho(x,t)$, behaves differently. For example, if $\rho = 0$ this Neumann problem has the explicit Fourier development

$$u_0(x,t) = g_0 + \sum_{n=1}^{\infty} g_n e^{-\lambda_n^2 t} \cos(n\pi x) = \frac{1}{2} - \frac{1}{2} e^{-\pi^2 t} \cos(\pi x).$$

The computed solution, which is shown in Figure 6.2, is analytically compared to this Fourier development.

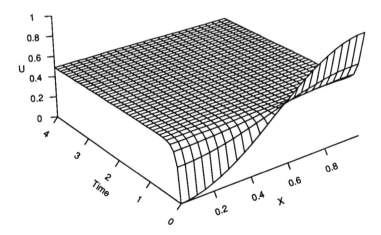

Figure 6.2 Computed solution with $M = 16$ for the Neumann problem (6.57) with $\rho = 0$.

For $\rho > 0$, the solution of the problem is given by

$$u_\rho(x,t) = \sum_{n=1}^{\infty} g_n e^{-\mu_n^2 t} \zeta_n(x)$$

where the eigenvalues μ_n satisfy the transcendental equation

$$\tan(\mu_n) = \frac{\rho}{\mu_n}$$

and the coefficients g_n are given by

$$g_n = \frac{2\mu_n \sin(\mu_n)}{\rho - \mu_n \sin^2(\mu_n)} \int_0^1 g(x)\zeta_n(x)dx.$$

The eigenfunctions are given by

$$\zeta_n(x) = \cos\left(\mu_n(1-x)\right).$$

The computed solutions for $\rho = 1$ and 10 are shown in Figures 6.3 and 6.4, respectively, where $M = 16$.

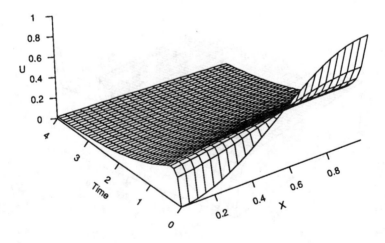

Figure 6.3 Computed solution with $M = 16$ for (6.57) with $\rho = 1$.

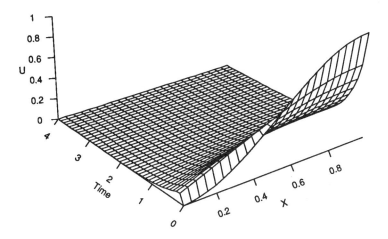

Figure 6.4 Computed solution with $M = 16$ for (6.57) with $\rho = 10$.

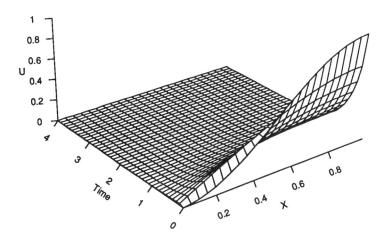

Figure 6.5 Computed solution with $M = 16$ for (6.57) with $\rho = \infty$.

The computed solution in the case that $\rho = \infty$ is shown in Figure 6.5. Since the boundary condition at $x = 0$ is $u(0,t) = 0$, the true solution is

$$u_\infty(x,t) = \sum_{n=1}^\infty g_n e^{-\left(\frac{(2n-1)\pi}{2}\right)^2 t} \sin\left(\frac{(2n-1)\pi}{2}x\right).$$

where

$$g_n = 2\int_0^1 g(x) \sin\left(\frac{(2n-1)\pi}{2}x\right) dx.$$

For some analytic comparisons Table 6.4 displays the computed approximate solutions to $u_\rho(x,t)$ at the time levels $t = .2, .5, 1, 2$, and 4 for the different values of the parameter $\rho = 0, 1, 10$, and ∞. Reading from left to right for $\rho = 0$ shows the approximate solution approaching $1/2$, the steady state if $\rho = 0$. For $\rho > 0$, as t increases, the approximate solution approaches the zero steady state. This limiting behavior is ρ dependent since $u_\rho(x,t)$ approaches zero more rapidly as ρ increases. This will occur since for increasing ρ, the μ_n increase to their limiting value $(2n-1)\pi/2$.

ρ	$t = .2$	$t = .5$	$t = 1$	$t = 2$	$t = 4$
0	0.5669	0.5011	0.5019	0.5007	0.5001
1	0.5574	0.4154	0.2855	0.1357	0.0307
10	0.5314	0.2822	0.0992	0.0114	0.0001
∞	0.5170	0.2425	0.0678	0.0042	0.0000

Table 6.4 Values of the computed solution with $M = 16$ to (6.5) at several time levels for $\rho = 0, 1, 10$, and ∞.

■

The transport problem

$$\begin{aligned}
L_c u(x,t) &\equiv \frac{\partial u}{\partial t}(x,t) + \kappa \frac{\partial u}{\partial x}(x,t) - \frac{\partial^2 u}{\partial x^2}(x,t) = f(x,t) \\
u(a,t) &= \gamma(t) \quad u(b,t) = \delta(t), \quad t \geq 0 \\
u(x,0) &= g(x), \quad a \leq x \leq b
\end{aligned} \qquad (6.58)$$

TIME-DEPENDENT PROBLEMS

as well as other advection-diffusion models can also be handled by the methods of this section with a simple adjustment in the spatial matrix to account for the first derivative term appearing in (6.58). The development of the method, as in the case of the nonhomogeneous heat problem, is most directly obtained by first transforming the problem to a homogeneous problem. This is done in exactly the same way as it is done for the heat equation. That is, use the sequence of transformations in (6.33) through (6.36). Hence, the obligation here is the introduction of the discretization of the first derivative term in (6.58) to the system (6.6). Actually this discretization has already been carried out, once for the first-order problem in (4.77) and again at the beginning of this chapter for the temporal discretization giving B_t in (6.10). For the same reasons that the weight in Example 4.6 was taken to be $w(x) = 1/\phi'(x)$, this selection is used here. Recalling the definition of $\mathcal{B}(1/\phi')$ ((4.79) with $w = 1/\phi'$), shows that $\kappa D(\phi')\mathcal{B}(1/\phi')$ needs to be added to the spatial matrix in (6.6). Hence, the Sinc-Galerkin discretization for the spatial derivatives in (6.58) is given by

$$\mathcal{A}_C = D(\phi')\mathcal{A}\left(\frac{1}{\phi'}\right)D(\phi') + \kappa D(\phi')\mathcal{B}\left(\frac{1}{\phi'}\right). \qquad (6.59)$$

The convection matrix in (6.59) is a diagonal multiple (the left multiplication by $D(\phi')$) of the coefficient matrix in the scalar convection Example 4.6. Introducing this change of spatial weight in (6.8) defines the weight function used for the problem (6.58) by

$$w(x)\tau(t) = \frac{\sqrt{\Upsilon'(t)}}{\phi'(x)}.$$

Combining this spatial matrix with the temporal matrix B_t in (6.10), gives the discrete system for the problem (6.58):

$$\mathcal{A}_C V^{(2)} + V^{(2)} B_t^T = G^{(2)}. \qquad (6.60)$$

The transformed matrix of unknowns is, from (6.7),

$$V^{(2)} = D\left(\frac{1}{\phi'}\right) U^{(2)}$$

and the matrix $G^{(2)}$ has exactly the same meaning as it did in (6.7), where the transformed forcing term $\bar{f}$ has the same meaning with the operator $P^{(2)}$ in (6.36) replaced by L_c in (6.58).

Example 6.6 The problem

$$L_c u(x,t) \equiv \frac{\partial u}{\partial t}(x,t) + \kappa \frac{\partial u}{\partial x}(x,t) - \frac{\partial^2 u}{\partial x^2}(x,t) = f(x,t)$$
$$u(0,t) = \gamma(t) = (t^2+1)e^{-(1+\frac{1}{\kappa})t}, \quad t \geq 0 \qquad (6.61)$$
$$u(1,t) = \delta(t) = 0, \quad t \geq 0$$
$$u(x,0) = g(x) = \sin(\pi x) + (1-x), \quad 0 \leq x \leq 1$$

is in a form that is directly amenable to the procedure outlined above. The system (6.60) is solved for the coefficients in (6.35). The function $\bar{f}$ in (6.36) takes the form

$$\bar{f}(x,t) = f(x,t) - L_c\left[\bar{\gamma}(t)(1-x) + g(x)\frac{t+1}{t^2+1}\right]$$

since $\bar{\delta}(t) = \delta(t) = 0$ ($g(1) = 0$). Further, $g(0) = 1$ so that $\bar{\gamma}(t) = \gamma(t) - ((t+1)/(t^2+1))$. The true solution of the problem is given by

$$u(x,t) = \left(t^2+1\right)e^{-(1+\frac{1}{\kappa})t}\left(\sin(\pi x) + (1-x)\right).$$

The results for $\kappa = 100$ are given in Table 6.5, where the quantities listed under the heading $\|E_S(h_s)\|$ are, as usual, the error on the sinc grid.

M_x	N_x	M_t	N_e	$h = \pi/\sqrt{2M_x}$	$\|E_S(h_s)\|$
4	4	4	2	1.111	$.632 \times 10^{-1}$
8	8	8	3	0.785	$.463 \times 10^{-2}$
16	16	16	4	0.555	$.169 \times 10^{-3}$
32	32	32	7	0.393	$.469 \times 10^{-5}$

Table 6.5 The solution of the advection-diffusion problem (6.61) with $\kappa = 100$.

Following the discussion in the nonhomogeneous Example 4.16, the parameter choices are $\alpha = \beta = \gamma = 1$ and $\delta = 1 + 1/\kappa \approx 1$.

TIME-DEPENDENT PROBLEMS   257

Thus $h = \pi/\sqrt{2M_x}$ ($d = \pi/2$) and $M_x = N_x = M_t$. Further, due to exponential decay,

$$N_e = \left[\left|\frac{\sqrt{2M_x}}{\pi}\ell n\left(\pi\sqrt{\frac{M_x}{2}}\right) + 1\right|\right].$$

∎

6.3 Burgers' Equation

A nonlinear advection-diffusion equation (Burgers' equation) is defined by

$$\begin{aligned}\frac{\partial u}{\partial t}(x,t) &+ \frac{\partial u}{\partial x}(x,t) - \epsilon\frac{\partial^2 u}{\partial x^2}(x,t) = f(x,t)\\ u(0,t) &= \gamma(t)\ u(1,t) = \delta(t),\ t \geq 0\\ u(x,0) &= 0,\ 0 \leq x \leq 1.\end{aligned} \qquad (6.62)$$

The Sinc-Galerkin discretization of the differential equation is, for all but the nonlinearity, exactly the same as in the previous linear transport problem. The nonlinearity $u\partial u/\partial x$ can be approached in a couple of fashions (all leading to the same discretization), and perhaps the simplest is first to write this term as $(1/2)\partial(u^2)/\partial x$.

An inspection of the system in (4.78) (just replace u by u^2) shows that the coefficient matrix of a sinc discretization of the nonlinear term $\partial(u^2)/\partial x$ in (6.62) is given by

$$\mathcal{A}_B(w) = \left[-\frac{1}{h}I^{(1)} - D\left(\frac{w'}{\phi'w}\right)\right]D(w). \qquad (6.63)$$

In terms of defined quantities, $\mathcal{A}_B(w)D(1/w) = \mathcal{B}(w)$, where $\mathcal{B}(w)$ is the matrix defined in (4.79). It is convenient in the following discussion to isolate the matrix $\mathcal{A}_B(w)$.

Incorporating the additional term involving the coefficient matrix in (6.63) into the system in (6.6) yields the discrete system for (6.62)

$$\epsilon D(\phi')\mathcal{A}(w)D(\phi')V^{(2)} + V^{(2)}B_t^T \qquad (6.64)$$

$$= G^{(2)} - \frac{1}{2}\mathcal{A}_B(w)V^{(2)} \circ V^{(2)}.$$

The notation $V^{(2)} \circ V^{(2)}$ denotes the Hadamard, or element-by-element, product of $V^{(2)}$ with itself (Definition A.25). The left-hand side of (6.64) is the same as the left-hand side of (6.6). The matrix B_t is given in (6.10) and $\mathcal{A}(w)$ is given in (4.40). The matrices $V^{(2)} = D(w)U^{(2)}$ and $G^{(2)} = D(w)F^{(2)}$ are exactly as in (6.7).

The choice of spatial weight w is similar to that for the linear advection-diffusion equation that led to the discretization in (6.60). Indeed, if $w(x) = 1/\phi'(x)$ then the system in (6.64) is

$$\epsilon D(\phi')\mathcal{A}\left(\frac{1}{\phi'}\right) D(\phi')V^{(2)} + V^{(2)}B_t^T$$

$$= G^{(2)} - \frac{1}{2}\mathcal{A}_B\left(\frac{1}{\phi'}\right) V^{(2)} \circ V^{(2)}. \tag{6.65}$$

where

$$\mathcal{A}_B\left(\frac{1}{\phi'}\right) = \left[-\frac{1}{h}I^{(1)} + D(2x-1)\right] D(x(1-x)). \tag{6.66}$$

Before implementing the above weight selection, there is reason to believe in the method presented in this section which is motivated by the selection of the spatial weight

$$w(x) = \frac{1}{\sqrt{\phi'(x)}}.$$

When substituted in (6.64), this selection gives the system

$$\epsilon A_x V^{(2)} + V^{(2)}B_t^T = G^{(2)} - \frac{1}{2}\mathcal{A}_B\left(\frac{1}{\sqrt{\phi'}}\right) V^{(2)} \circ V^{(2)} \tag{6.67}$$

where

$$\mathcal{A}_B\left(\frac{1}{\sqrt{\phi'}}\right) = \left[-\frac{1}{h}I^{(1)} + D\left(\frac{2x-1}{2}\right)\right] D\left(\sqrt{x(1-x)}\right). \tag{6.68}$$

There are two methods that come to mind for the solution of (6.67). The first would involve casting this problem as a nonlinear system and using a nonlinear solver such as Newton's method [9]. In an attempt to keep matters simple, an alternative procedure that is

TIME-DEPENDENT PROBLEMS

motivated by the spectral radius bound in (6.29) is indicated. Concatenating the system in (6.67) gives

$$(I_{m_t} \otimes \epsilon A_x + B_t \otimes I_{m_x}) \operatorname{co}\left(V^{(2)}\right)$$
$$= \operatorname{co}\left(G^{(2)}\right) - \frac{1}{2}\left(I_{m_t} \otimes \mathcal{A}_B\left(\frac{1}{\sqrt{\phi'}}\right)\right) \operatorname{co}\left(V^{(2)} \circ V^{(2)}\right). \tag{6.69}$$

By (6.29) the spectral radius of the inverse of the coefficient matrix on the left is bounded above by $(b-a)^2/(4\epsilon) = 1/(4\epsilon)$. If this bound is very conservative, then an iterative scheme is suggested by (6.69). Whether the bound is conservative or not it is the case that, if ϵ is not too large, then the iterative method

$$(I_{m_t} \otimes \epsilon A_x + B_t \otimes I_{m_x}) \operatorname{co}\left(V_n^{(2)}\right)$$
$$= \operatorname{co}\left(G^{(2)}\right) - \frac{1}{2}\left(I_{m_t} \otimes \mathcal{A}_B\left(\frac{1}{\sqrt{\phi'}}\right)\right) \operatorname{co}\left(V_{n-1}^{(2)} \circ V_{n-1}^{(2)}\right) \tag{6.70}$$

can be used to compute the solution to (6.62).

The calculations requested in Exercise 6.4 give, for a specific example, limitations of this iterative procedure. Although the results of Exercise 6.4 do not bring one closer to bounding the spectral radius of the inverse matrix in (6.69), they do give some positive indications concerning this iterative approach. However, as ϵ decreases the iterative method deteriorates (diverges) and nonlinear solvers may be used to obtain calculations for smaller values of ϵ. This too will eventually deteriorate, and methods using conservation law considerations with extrapolation will perhaps provide a more useful line of attack ([14] and [15]). The recent extrapolation procedures discussed in [4] may well play a role.

The best attack for the problem in (6.62) is far from settled, and the idea of a sort of regularization of the iterative procedure, implemented via adjusting the interplay between the mesh size h and the parameter ϵ, while distinct from more standard regularization procedures, may not be too farfetched. At least for moderately small values of ϵ the system in (6.65) handles, via the iteration scheme in (6.70), the calculation of the solution of (6.62) quite well.

Example 6.7 Burgers' system defined by

$$\frac{\partial u}{\partial t}(x,t) + u(x,t)\frac{\partial u}{\partial x}(x,t) = \epsilon \frac{\partial^2 u}{\partial x^2}(x,t)$$
$$u_x(0,t) = \rho u(0,t), \quad t \geq 0 \quad (6.71)$$
$$u_x(1,t) = 0, \quad t \geq 0$$
$$u(x,0) = (x(1-x))^2, \quad 0 \leq x \leq 1$$

is a sort of nonlinear version of the problem worked out in Example 6.5. The same spatial weight $w(x) = 1/\phi'(x)$ is used in this example. The total discretization of this problem from scratch is a lengthy process, but the preceding discussion should lend plausibility to the following development. The nonhomogeneous initial condition is subtracted from the system using the change of variable

$$w(x,t) = u(x,t) - \theta(t)g(x).$$

This transforms (6.71) into a problem with homogeneous initial data (Exercise 6.3). The differentiable function $\theta(t)$ is chosen to satisfy $\theta(0) = 1$ (see Exercise 6.4). Actually, putting $g(x) = (x(1-x))^2$ in Exercise 6.4 gives the function $\bar{f}(x,t)$ for this example. The problem that results is then one with homogeneous radiation boundary conditions and, except for the nonlinearity, the problem is in the form of (6.39). Augmenting the sinc basis in exactly the same fashion as was done in (6.41) and (6.42), leads to the bordered matrices in (6.51) through (6.54). Each of these bordered matrices occurs in the system in (6.65). All that is left to complete the system is to note that the matrix appearing in (6.66) has size $(m_x + 2) \times (m_x + 2)$. It is not bordered because the orthogonalization has been carried out with respect to two extra spatial sinc basis functions ($p = -M_x - 1$ and $N_x + 1$) in (6.43). Piecing this together yields the iterative scheme

$$\left(\mathcal{D}_t^T \otimes \epsilon \mathcal{A}\left(\frac{1}{\phi'}\right) + \mathcal{B}^T \otimes \mathcal{D}_s\right) \operatorname{co}\left(C_n^{(2)}\right)$$

$$= \operatorname{co}(\mathcal{F}) - \left(I_{m_t} \otimes \mathcal{A}_B\left(\frac{1}{\phi'}\right)\right) \operatorname{co}\left(C_{n-1}^{(2)} \circ C_{n-1}^{(2)}\right) \quad (6.72)$$

to determine the matrix $C_n^{(2)}$. The spatial bordered matrices $\mathcal{A}(1/\phi')$ and $\mathcal{D}_s$ have exactly the same definition as in (6.51) and (6.52), respectively. The spatial temporal matrices $\mathcal{B}$ and $\mathcal{D}_t$ are given in

(6.53) and (6.54), respectively. The matrix $C_n^{(2)}$ contains the coefficients in the assumed approximate solution (6.45). The particular form of the two functions in (6.41) required to fill the borders is

$$\omega_0(x) = [(2\rho+1)x + \rho](1-x)^2$$
$$\omega_1(x) = (3-2x)x^2.$$

The function

$$\omega_\infty(t) = \frac{t}{t+1}$$

is the same as in (6.42).

Beginning with the zero initial matrix, the iteration in (6.72) is run until the error between successive iterates is less than 10^{-4}. The parameters to define the iteration, in keeping with the discussion in Example 4.6, are $\alpha = \beta = \gamma = 1$ and $\delta = 1$. Thus $h = \pi/\sqrt{2M_x}$ $(d = \pi/2)$ and $M \equiv M_x = N_x = M_t = N_t$. The results with $\epsilon = 0.01$ are shown in Figures 6.6, 6.7, 6.8, and 6.9 for $\rho = 0, 1, 10, \infty$, respectively. The number of iterations used to generate the results is listed in each caption as n_f. All systems were solved with $M = 8$.

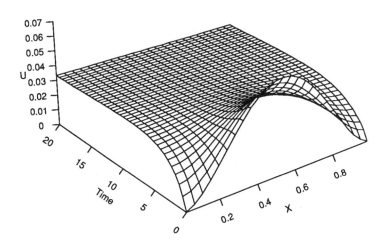

Figure 6.6 Fully Sinc-Galerkin solution to (6.71) with $M = 8$ and Neumann boundary conditions ($n_f = 4$).

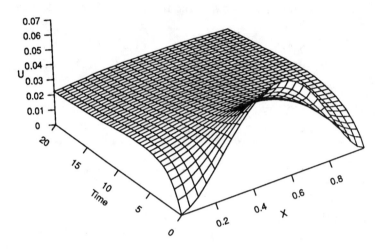

Figure 6.7 Fully Sinc-Galerkin solution with $M = 8$ to (6.71) with $\rho = 1$ ($n_f = 3$).

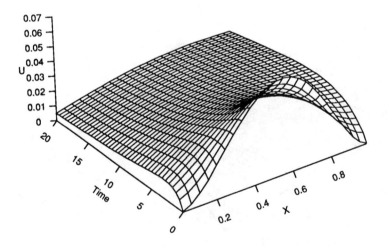

Figure 6.8 Fully Sinc-Galerkin solution with $M = 8$ to (6.71) with $\rho = 10$ ($n_f = 3$).

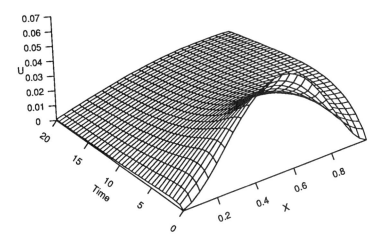

Figure 6.9 Fully Sinc-Galerkin solution with $M = 8$ to (6.71) with $\rho = \infty$ ($n_f = 3$).

■

Exercises

Exercise 6.1 The problem

$$P^{(2)}u(x,t) \equiv \frac{\partial u}{\partial t}(x,t) - \frac{\partial^2 u}{\partial x^2}(x,t) = f(x,t)$$
$$u(0,t) = u(1,t) = 0, \quad t \geq 0 \qquad (6.73)$$
$$u(x,0) = 0, \quad 0 \leq x \leq 1$$

has the rational (in the temporal domain) solution

$$u(x,t) = \frac{x(1-x)t}{1+t^2}$$

when

$$f(x,t) = (1 + 2t - t^2 + 2t^3)\frac{x(1-x)}{(t^2+1)^2}.$$

The discrete system for this problem is given by (6.9) and, except for $F^{(2)}$, consists of the matrices listed in Example 6.2. The solution $U^{(2)}$ gives the coefficients in the approximate Sinc-Galerkin solution

$$u_{m_x,m_t}(x,t) = \sum_{j=-M_t}^{N_t} \sum_{i=-M_x}^{N_x} u_{ij} S_{ij}(x,t)$$

where

$$S_{ij}(x,t) \equiv S(i,h_s) \circ \left(\ell n \left(\frac{x}{1-x} \right) \right) S(j,h_s) \circ (\ell n(t)).$$

From (6.12), $\alpha_s = \beta_s = \gamma_s = \delta_s = 1/2$ so $M_x = M_t = N_x$ and $h_s = \pi/\sqrt{M_x}$ ($d = \pi/2$). Calculate the approximate solution u_{m_x,m_t} using the values $M_x = 4, 8, 16,$ and 32 with N_e chosen from (6.22). Repeat this calculation using $N_t = M_x$. Recalling the discussion in Examples 4.10 and 4.13, explain why one cannot always economize the discrete system.

Exercise 6.2 Assume λ_t is in the spectrum of

$$S = D((\Upsilon')^{1/2}) \left[-\frac{1}{h} I^{(1)} - D\left(\frac{\tau'}{\tau \Upsilon'}\right) \right] D((\Upsilon')^{1/2}).$$

If $\Upsilon(t) = \ell n(\sinh(t))$, show that

$$\frac{\tau'(t)}{\tau(t)\Upsilon'(t)} = \frac{-1}{2\cosh^2(t)}$$

where the weight $\tau(t) = \sqrt{\Upsilon'(t)}$. Conclude that $Re(\lambda_t) > 0$. Does the bound in (6.29) hold for λ, where $\lambda = \lambda_x + \lambda_t$ and λ is in the spectrum of

$$\mathcal{B}_s^{(2)} = I_{m_t} \otimes A_x + S \otimes I_{m_x}$$

where, recalling (5.89),

$$A_x \equiv D(\phi'_x) \left[-\frac{1}{h_x^2} I^{(2)} + D\left(\frac{1}{4}\right) \right] D(\phi'_x)?$$

Compare this with the discussion preceding (6.29).

Exercise 6.3 Consider the transformation
$$w(x,t) = u(x,t) - \theta(t)g(x), \tag{6.74}$$
where the differentiable function $\theta(t)$ satisfies $\theta(0) = 1$. In particular, as in (6.32), take $\theta(t) = \theta_{-M_t-1}(t)$, that is
$$\theta(t) = \frac{t+1}{t^2+1}.$$
The transformation (6.74) converts the problem with nonhomogeneous initial data g;
$$\begin{aligned} P^{(2)}u(x,t) &\equiv \frac{\partial u}{\partial t}(x,t) - \frac{\partial^2 u}{\partial x^2}(x,t) = f(x,t) \\ a_1 u_x(0,t) &- a_0 u(0,t) = 0, \quad t \geq 0 \\ b_1 u_x(1,t) &+ b_0 u(1,t) = 0, \quad t \geq 0 \\ u(x,0) &= g(x), \quad 0 \leq x \leq 1, \end{aligned} \tag{6.75}$$
into the problem with homogeneous initial data
$$\begin{aligned} P^{(2)}w(x,t) &\equiv \frac{\partial w}{\partial t}(x,t) - \frac{\partial^2 w}{\partial x^2}(x,t) = \bar{f}(x,t) \\ a_1 w_x(0,t) &- a_0 w(0,t) = 0, \quad t \geq 0 \\ b_1 w_x(1,t) &+ b_0 w(1,t) = 0, \quad t \geq 0 \\ w(x,0) &= 0, \quad 0 \leq x \leq 1 \end{aligned} \tag{6.76}$$
and
$$\bar{f}(x,t) = f(x,t) + \theta(t)\frac{d^2}{dx^2}(g(x)) - g(x)\frac{d}{dt}(\theta(t)).$$
For the solution to be continuous at $(0,0)$ and $(1,0)$, the initial data $g(x)$ satisfies the boundary conditions. If $a_1 = b_1 = 0$ in (6.76), then the problem is the homogeneous ($\gamma(t) = \delta(t) = 0$) Dirichlet problem in (6.30) on $(0,1)$. Show that $\bar{f}(x,t)$ defined above is the same function as defined in (6.36).

Exercise 6.4 A classic Burgers' problem with homogeneous boundary conditions is defined by
$$\begin{aligned} \frac{\partial u}{\partial t}(x,t) &+ u(x,t)\frac{\partial u}{\partial x}(x,t) = \epsilon\frac{\partial^2 u}{\partial x^2}(x,t) \\ u(0,t) &= 0, \quad t \geq 0 \\ u(1,t) &= 0, \quad t \geq 0 \\ u(x,0) &= \sin(\pi x), \quad 0 \leq x \leq 1. \end{aligned} \tag{6.77}$$

The change of variables defined in Exercise 6.3 transforms (6.77) into the system with homogeneous initial data

$$\frac{\partial w}{\partial t}(x,t) + w(x,t)\frac{\partial w}{\partial x}(x,t) + \theta(t)g(x)\frac{\partial w}{\partial x}(x,t)$$

$$+\theta(t)\frac{d}{dx}(g(x))w(x,t) = \epsilon\frac{\partial^2 w}{\partial x^2}(x,t) + \bar{f}(x,t)$$

$$w(0,t) = 0, \quad t \geq 0$$
$$w(1,t) = 0, \quad t \geq 0$$
$$w(x,0) = 0, \quad 0 \leq x \leq 1.$$

A bit of patience shows that the general form for the function $\bar{f}(x,t)$ is

$$\bar{f}(x,t) = \epsilon\theta(t)\frac{d^2}{dx^2}(g(x)) - g(x)\frac{d}{dt}(\theta(t)) - \theta^2(t)g(x)\frac{d}{dx}(g(x)).$$

One could use the function $\theta(t)$ defined in Exercise 6.3 but, as mentioned in [1], there are analytic reasons that motivate the selection $\theta(t) = \exp(-t)$. The resulting function $\bar{f}(x,t)$ takes the form

$$\bar{f}(x,t) = \left(\epsilon\frac{d^2}{dx^2}(g(x)) + g(x) - e^{-t}g(x)\frac{d}{dx}(g(x))\right)e^{-t}.$$

Select $\alpha = \beta = \gamma = \delta = 1$ so that $h = \pi/\sqrt{2M_x}$ ($d = \pi/2$) and $M_x = N_x = M_t$ with

$$N_e = \left[\left|\frac{\sqrt{2M_x}}{\pi}\ell n\left(\pi\sqrt{\frac{M_x}{2}}\right) + 1\right|\right]$$

and implement the iteration in (6.69) for $M_x = 16$, using the initial guess $V_0^{(2)} = 0$. Denote the errors between the computed iterates by

$$E^{(n)} = \max_{\substack{-M_x \leq j \leq N_x \\ -M_t \leq k \leq N_t}} |u_{jk}^{(n)} - u_{jk}^{(n-1)}|$$

where the matrix of coefficients in the assumed approximate solution $U_n^{(2)}$ in (6.2) is obtained by transforming back with $U_n^{(2)} = D(\phi')V_n^{(2)}$. Define the ratio $ICR(n) = E^{(n)}/E^{(n-1)}$, and truncate the algorithm when $ICR(n)$ is approximately constant. Run the algorithm for $\epsilon = 1, .1, .05, .01,$ and $.005$.

References

[1] J. D. Cole, "On a Quasi-Linear Parabolic Equation Occurring in Aerodynamics," *Quarterly Appl. Math.*, 9 (1951), pages 225–236.

[2] R. R. Doyle, "Extensions to the Development of the Sinc-Galerkin Method for Parabolic Problems," Ph.D. Thesis, Montana State University, Bozeman, MT, 1990.

[3] G. H. Golub and C. F. Van Loan, *Matrix Computations*, Second Ed., Johns Hopkins University Press, Baltimore, 1989.

[4] S. - Å. Gustafson and F. Stenger, "Convergence Acceleration Applied to Sinc Approximation with Application to Approximation of $|x|^\alpha$," in *Computation and Control II*, Proc. Bozeman Conf. 1990, Progress in Systems and Control Theory, Vol. 11, Birkhäuser, Boston, 1991, pages 161–171.

[5] D. L. Lewis, "A Fully Galerkin Method for Parabolic Problems," Ph. D. Thesis, Montana State University, Bozeman, MT, 1989.

[6] D. L. Lewis, J. Lund, and K. L. Bowers, "The Space-Time Sinc-Galerkin Method for Parabolic Problems," *Internat. J. Numer. Methods Engrg.*, 24 (1987), pages 1629–1644.

[7] J. Lund, K. L. Bowers, and T. S. Carlson, "Fully Sinc-Galerkin Computation for Boundary Feedback Stabilization," *J. Math. Systems, Estimation, and Control*, 1 (1991), pages 165–182.

[8] J. Lund, K. L. Bowers, and K. M. McArthur, "Symmetrization of the Sinc-Galerkin Method with Block Techniques for Elliptic Equations," *IMA J. Numer. Anal.*, 9 (1989), pages 29–46.

[9] L. Lundin, "A Cardinal Function Method of Solution of the Equation $\Delta u = u - u^3$," *Math. Comp.*, 35 (1980), pages 747–756.

[10] K. M. McArthur, "Sinc-Galerkin Solution of Second-Order Hyperbolic Problems in Multiple Space Dimensions," Ph. D. Thesis, Montana State University, Bozeman, MT, 1987.

[11] K. M. McArthur, K. L. Bowers, and J. Lund, "Numerical Implementation of the Sinc-Galerkin Method for Second-Order Hyperbolic Equations," *Numer. Methods Partial Differential Equations*, 3 (1987), pages 169–185.

[12] K. M. McArthur, K. L. Bowers, and J. Lund, "The Sinc Method in Multiple Space Dimensions: Model Problems," *Numer. Math.*, 56 (1990), pages 789–816.

[13] R. C. Smith, K. L. Bowers, and J. Lund, "A Fully Sinc-Galerkin Method for Euler–Bernoulli Beam Models," *Numer. Methods Partial Differential Equations*, 8 (1992), pages 171–202.

[14] M. Stromberg, "Solution of Shock Problems by Methods Using Sinc Functions", Ph. D. Thesis, University of Utah, Salt Lake City, UT, 1988.

[15] M. Stromberg, "Sinc Approximate Solution of Quasilinear Equations of Conservation Law Type," in *Computation and Control*, Proc. Bozeman Conf. 1988, Progress in Systems and Control Theory, Vol. 1, Birkhäuser, Boston, 1989, pages 317–331.

Appendix A

Linear Algebra

This appendix presents material from the theory of linear algebra that has occurred at various points throughout the text. Most of the material is covered in courses in matrix theory, but it has proved useful to have in hand the results germane to the text development. The particular results concerning Toeplitz matrices seem less commonly known (and appreciated) than the topic deserves. The section on block matrices and tensor products, with the unraveling operators "mat" and "co," is a valuable tool in the cataloguing (if not the solution method) of the discretizations of partial differential equations. Most of the material is found in a number of sources such as [1], [3], [5], [6], and [7].

A.1 Special Matrices and Diagonalization

In the discretization of boundary value problems a differential equation is replaced with a linear system of equations. Various questions naturally arise about the properties of the system. Often these questions can be answered by the spectral structure of the coefficient matrix of the system. There are various tools from matrix theory that are useful in addressing these questions. Much of the information concerning the theory and computation of eigenvalues that is presented here is found in [6] and [7]. This section summarizes a number of the tools from linear algebra that will be helpful in analyzing the matrices that occurred in the text.

Throughout the appendix, capital letters (A, B, C, ...) will denote matrices. If the elements of A need specification, then lowercase letters (a_{ij}) will be used to denote the entries. Column vectors are represented by lowercase letters and an arrow ($\vec{u}$, $\vec{v}$, $\vec{w}$, ...). If $f(x)$ is a function defined on the set $\{x_i\}_{i=1}^n$, then the $n \times n$ diagonal matrix with diagonal entries $d_{ii} = f(x_i)$, $1 \leq i \leq n$, will be denoted by $D(f)$.

The appropriate setting for most of what follows is $\mathbb{C}^n$, which is a normed linear space with norm

$$\|\vec{z}\|_p \equiv \left\{\sum_{i=1}^n |z_i|^p\right\}^{1/p}, \quad p \geq 1.$$

Letting $p \to \infty$ gives the maximum norm

$$\|\vec{z}\|_\infty \equiv \max_{1 \leq i \leq n} |z_i|.$$

As with the $L^p(a, b)$ spaces, a frequently occurring case is when $p = 2$ and that norm (called the *Euclidean norm*) arises from the inner product

$$(\vec{z}, \vec{w}) = \sum_{i=1}^n \bar{z}_i w_i. \tag{A.1}$$

Several special matrices that have distinguishing properties are included in the following definition.

Definition A.1 Let A be an $n \times n$ matrix. The *transpose* of A is $A^T = (a_{ji})$. The *conjugate transpose* $\bar{A}^T = (\overline{a_{ji}})$ is called the *adjoint* of A and is denoted $A^* \equiv \bar{A}^T$. A matrix A is said to be *Hermitian* if $A^* = A$ and *skew-Hermitian* if $A^* = -A$. The real matrix A that satisfies $A^T = A$ (or $A^T = -A$) is called *symmetric* (or *skew-symmetric*). A class of matrices that subsumes both of the preceding types is the set of *normal* matrices for which $AA^* = A^*A$.

With the notation just introduced the inner product in (A.1) can be written

$$(\vec{z}, \vec{w}) = \vec{z}^* \vec{w}.$$

Definition A.2 If A is an $n \times n$ matrix and $\vec{v}$ is a nonzero column vector satisfying the equation

$$A\vec{v} = \lambda \vec{v} \qquad (A.2)$$

for some number $\lambda \in \mathbb{C}$, then the pair $\{\lambda, \vec{v}\}$ is called an *eigenpair* of the matrix A. The vector $\vec{v}$ is called an *eigenvector* of A and λ is called an *eigenvalue* of A. The set of all distinct λ satisfying (A.2) for some $\vec{v}$ is the *spectrum* of A denoted by $\sigma(A)$. The eigenvalue with maximum modulus is called the *spectral radius* of A

$$\rho(A) = \max_{\lambda \in \sigma(A)} |\lambda|.$$

Definition A.3 A matrix Q satisfying

$$Q^*Q = I \qquad (A.3)$$

is called a *unitary* matrix (or an *orthogonal* matrix if Q is real). If there is a Q satisfying (A.3) and

$$Q^*AQ = D(\lambda), \qquad (A.4)$$

then the matrix A is said to be *unitarily similar to a diagonal matrix* (or *unitarily diagonalizable*). Again if Q is real, A is said to be *orthogonally similar to a diagonal matrix* or *orthogonally diagonalizable*.

The next two theorems classify the orthogonally diagonalizable real matrices and the unitarily diagonalizable complex matrices.

Theorem A.4 *The following are equivalent for an $n \times n$ real matrix A:*

(i) *A is orthogonally diagonalizable.*

(ii) *A has a full orthonormal set of eigenvectors.*

(iii) *A is symmetric.*

∎

Theorem A.5 *The following are equivalent for an $n \times n$ complex matrix A:*

(i) *A is unitarily diagonalizable.*

(ii) *A has a full orthonormal set of eigenvectors.*

(iii) *A is normal.*

∎

In particular, Hermitian and skew-Hermitian matrices are unitarily diagonalizable. There are matrices that are not normal, for which a diagonalization is possible (not, of course, a unitary one). Any matrix that is similar to a diagonal matrix via

$$P^{-1}AP = D(\lambda) \qquad (A.5)$$

is called *diagonalizable*. Note that P must contain the eigenvectors of A. Indeed, the matrix A is diagonalizable if and only if A has n linearly independent eigenvectors.

By Theorem A.5, if A is not normal it cannot be unitarily similar to a diagonal matrix. The answer to how far an arbitrary matrix can be unitarily reduced is given in the following theorem.

Theorem A.6 (Schur's Theorem) *Any square matrix is unitarily similar to an upper triangular matrix.*

∎

Hermitian and skew-Hermitian matrices have a special spectral structure. This is described next.

Theorem A.7 *If A is Hermitian, then $\sigma(A) \subset \mathbb{R}$. In fact, a normal matrix A is Hermitian if and only if $\sigma(A) \subset \mathbb{R}$.*

∎

If A is an $n \times n$ Hermitian matrix, then the eigenvalues will always be ordered from smallest to largest: $\lambda_1 \leq \lambda_2 \leq \cdots \leq \lambda_n$. A consequence of the above is that if A is skew-Hermitian, then the matrix iA is Hermitian. This provides the following corollary.

Corollary A.8 *If A is skew-Hermitian, then $\sigma(A) \subset i\mathbb{R} = \{iy : y \in \mathbb{R}\}$. In fact, a normal matrix A is skew-Hermitian if and only if $\sigma(A) \subset i\mathbb{R}$.*

∎

Hermitian matrices whose spectrum is a subset of $\mathbb{R}^+$ (positive reals) have the following special property.

Definition A.9 *The real symmetric matrix A is positive definite if for all nonzero $\vec{v} \in \mathbb{R}^n$.*

$$\vec{v}^T A \vec{v} > 0. \tag{A.6}$$

Definition A.10 *The complex Hermitian matrix A is positive definite if for all nonzero $\vec{v} \in \mathbb{C}^n$.*

$$\vec{v}^* A \vec{v} > 0. \tag{A.7}$$

In either of the above definitions, if equality is allowed then the matrix A is called positive semi-definite. Positive definite matrices can be characterized in several ways. The following facts are useful in facilitating these characterizations.

Theorem A.11 (Triangular Factorization Theorem) *If the leading principal submatrices of A are all nonsingular, then unique normalized factors L, D, and U exist such that $A = LDU$. The matrix L is lower triangular, D is diagonal, and U is upper triangular. L and U are normalized with ones on the main diagonal.*

∎

The matrix A can be written as $A = (LD^{1/2})(LD^{1/2})^* \equiv C^*C$ when A is positive definite. This is called the *Cholesky factorization* of A.

Theorem A.12 *The following are equivalent for an $n \times n$ Hermitian matrix A:*

(i) *A is positive definite.*

(ii) *All the leading principal submatrices of A have positive determinants.*

(iii) *A has a unique Cholesky factorization $A = C^*C$, where C is upper triangular with positive diagonal entries.*

∎

One important application of positive definite matrices occurs when solving Sturm–Liouville problems. After a discretization a *generalized eigenvalue problem*

$$A\vec{v} = \lambda B\vec{v} \qquad (A.8)$$

results. If B is the identity matrix, then (A.8) is called the *standard eigenvalue problem*. Here λ is called an *eigenvalue* and $\vec{v}$ the corresponding *eigenvector* of the *matrix pair* (A, B). The matrix $A - \lambda B$ is also referred to as a *matrix pencil*. If B is nonsingular, then this may be written as

$$B^{-1}A\vec{v} = \lambda \vec{v}$$

which is a standard eigenvalue problem for $B^{-1}A$. Even in this case, for numerical considerations it may not be feasible to write (A.8) in this manner. For example, symmetry is lost even if A and B are symmetric. In the case in which B is singular, the theory is more difficult. In particular, A and B may have identical vectors in their null spaces, in which case any number λ is an eigenvalue. Also, there may be fewer than n eigenvalues. If this is the case, and A is nonsingular consider,

$$A^{-1}B\vec{v} = \lambda^{-1}\vec{v} \equiv \mu\vec{v}.$$

LINEAR ALGEBRA

The missing eigenvalues correspond to the zero eigenvalues of the pair (B, A). The missing eigenvalues of (A, B) are called *infinite eigenvalues* for this reason.

Symmetric matrices in the pair (A, B) do not necessarily assure real eigenvalues for (A.8). To guarantee this, the further assumption that B is positive definite is needed. Then by Theorem A.12, B has a Cholesky factorization written for convenience as $B = (C^{-1})^*C^{-1}$. Hence (A.8) yields

$$AC\vec{w} = \lambda(C^{-1})^*\vec{w}$$

where

$$C^{-1}\vec{v} = \vec{w}. \qquad (A.9)$$

Multiplication by (C^*) yields

$$C^*AC\vec{w} = \lambda\vec{w}$$

which is a standard eigenvalue problem with the same eigenvalues as (A.8) and eigenvectors related by (A.9). Since C^*AC is Hermitian, there exists a unitary Q so that

$$D(\lambda) = Q^*(C^*AC)Q = (CQ)^*A(CQ)$$

and setting $S = CQ$ yields

$$S^*BS = (CQ)^*B(CQ) = Q^*(C^*BC)Q = Q^*Q = I.$$

Theorem A.13 *Assume that A is Hermitian and B is positive definite. Then the eigenvalues of the generalized eigenvalue problem are all real and have the same sign as the standard eigenvalues of A. Further, there exists a matrix S (consisting of the eigenvectors $\vec{v}$ as its columns) that simultaneously diagonalizes A and B as follows:*

$$S^*AS = D(\lambda), \quad S^*BS = I. \qquad (A.10)$$

∎

There is a special orthogonal matrix defined by

$$J = \begin{bmatrix} & & & & 1 \\ & & & \cdot & \\ & & \cdot & & \\ & \cdot & & & \\ & 1 & & & \\ 1 & & & & \end{bmatrix} \qquad (A.11)$$

that serves to define another form of symmetry for a matrix A. The matrix J satisfies $J = J^{-1} = J^T$ and, due to its structure, is frequently referred to as the *counteridentity*. Note that left multiplication of the matrix A by J reverses the order of the rows of A and right multiplication reverses the order of the columns of A. Symmetry about the "center" of a matrix is less common than the standard symmetry, but it is also useful and is defined in terms of this counteridentity.

Definition A.14 The matrix A is called *centrosymmetric* (symmetric about its center) if $JAJ = A$. In component form this says $a_{ij} = a_{n+1-i, n+1-j}$.

Symmetry and centrosymmetry need not occur together. Matrices that are both symmetric and centrosymmetric have an especially nice structure.

Let A be an $n \times n$ real symmetric centrosymmetric matrix and, if n is even, write A as

$$A = \begin{bmatrix} B & C^T \\ C & JBJ \end{bmatrix}$$

where B and C are $n/2 \times n/2$ and $B^T = B$, $C^T = JCJ$. Multiplication of A by the orthogonal matrix

$$Q = \frac{1}{\sqrt{2}} \begin{bmatrix} I & I \\ -J & J \end{bmatrix}$$

yields

$$Q^T A Q = \begin{bmatrix} B - JC & O \\ O & B + JC \end{bmatrix}. \qquad (A.12)$$

If n is odd, then A may be written

$$A = \begin{bmatrix} B & \vec{x} & C^T \\ \vec{x}^T & q & \vec{x}^T J \\ C & J\vec{x} & JBJ \end{bmatrix}$$

where B and C are $(n-1)/2 \times (n-1)/2$, $\vec{x}$ is $p \times 1$, and $B^T = B$, $C^T = JCJ$. Define the orthogonal matrix

$$Q = \frac{1}{\sqrt{2}} \begin{bmatrix} I & \vec{0} & I \\ \vec{0}^T & \sqrt{2} & \vec{0}^T \\ -J & \vec{0} & J \end{bmatrix}$$

and a multiplication shows that A splits as

$$Q^T A Q = \begin{bmatrix} B - JC & \vec{0} & O \\ \vec{0}^T & q & \sqrt{2}\vec{x}^T \\ O & \sqrt{2}\vec{x} & B + JC \end{bmatrix}. \qquad (A.13)$$

Computationally symmetric, centrosymmetric matrices need relatively few elements to determine the entire matrix. Thus storage can be minimized. A special subclass of the symmetric, centrosymmetric matrices is the class of symmetric Toeplitz matrices.

Definition A.15 The $n \times n$ matrix A is *Toeplitz* if its elements satisfy

$$a_{ij} = a_{j-i},$$

for $1 \leq i, j \leq n$. Thus A has the form

$$A = \begin{bmatrix} a_0 & a_1 & a_2 & \cdots & a_{n-1} \\ a_{-1} & a_0 & a_1 & \cdots & a_{n-2} \\ \vdots & \ddots & \ddots & \ddots & \vdots \\ \vdots & \ddots & \ddots & \ddots & a_1 \\ a_{-n+1} & \cdots & \cdots & a_{-1} & a_0 \end{bmatrix}. \qquad (A.14)$$

It should be noted that if A is Toeplitz, so is A^T; furthermore, Toeplitz matrices are also "symmetric" about the main counterdiagonal. A matrix A that is symmetric about the main counterdiagonal is called *persymmetric* and is characterized by $JAJ = A^T$. Toeplitz matrices will arise in a natural setting in the next section.

A.2 Eigenvalue Bounds

In many cases the explicit determination of the eigenvalues of a given matrix is not possible. Although less satisfactory, eigenvalue bounds are often suitable to determine features of the matrix. This section includes several theorems that provide eigenvalue bounds that may be helpful in estimations relating to properties of a given matrix. The notation is from [6].

Theorem A.16 (Cauchy's Interlace or Separation Theorem)
Let A be an $n \times n$ Hermitian matrix with component blocks A_i ($i = 1, 2, 3$) as

$$A = \begin{bmatrix} A_1 & A_2^* \\ A_2 & A_3 \end{bmatrix}$$

where A_1 is $m \times m$, $m < n$. Let A have eigenvalues $\lambda_1 \leq \lambda_2 \leq \cdots \leq \lambda_n$ and A_1 have eigenvalues $\mu_1 \leq \mu_2 \leq \cdots \leq \mu_m$. For convenience, let

$$\mu_i = \left\{ \begin{array}{ll} -\infty & if\ i \leq 0 \\ +\infty & if\ i > m \end{array} \right\}, \quad \mu_{-i} = \left\{ \begin{array}{ll} -\infty & if\ i > m \\ +\infty & if\ i \leq 0 \end{array} \right\}.$$

Then for $j = 1, \ldots, m$

$$\lambda_j \leq \mu_j \leq \lambda_{j+n-m} \quad \text{and} \quad \lambda_{-(j+n-m)} \leq \mu_{-j} \leq \lambda_{-j}.$$

Alternatively, for $k = 1, \ldots, n$

$$\mu_{k-n+m} \leq \lambda_k \leq \mu_k \quad \text{and} \quad \mu_{-k} \leq \lambda_{-k} \leq \mu_{-(k-n+m)}.$$

■

The eigenvalues of the Hermitian matrix A can be characterized in terms of the *Rayleigh quotient*

$$R(\vec{x}) \equiv \frac{\vec{x}^* A \vec{x}}{\vec{x}^* \vec{x}}; \quad \vec{x} \neq \vec{0}. \tag{A.15}$$

In fact R is *stationary* at $\vec{x}$ (the gradient of R is zero at $\vec{x}$) if and only if $\vec{x}$ is an eigenvector of A. In particular the smallest and largest eigenvalues are $\lambda_1 = \min_{\vec{x} \neq \vec{0}} R(\vec{x})$ and $\lambda_n = \max_{\vec{x} \neq \vec{0}} R(\vec{x})$, respectively. Each λ_j can be characterized in terms of the eigenvectors $\vec{x}_1, \ldots, \vec{x}_{j-1}$ [5, page 287]. A more convenient characterization that avoids the use of the eigenvectors is the following theorem.

Theorem A.17 (Courant–Fischer Minimax Theorem) *Let C_j ($1 \leq j \leq n$) denote an arbitrary $(n - j + 1)$-dimensional subspace of $\mathbb{C}^n$ and let A be a Hermitian matrix with eigenvalues $\lambda_1 \leq \lambda_2 \leq \cdots \leq \lambda_n$. Then*

$$\lambda_j = \max_{C_j} \min_{\vec{0} \neq \vec{x} \in C_j} \frac{\vec{x}^* A \vec{x}}{\vec{x}^* \vec{x}}$$

or in a different form

$$\lambda_{n-j+1} = \min_{C_j} \max_{\vec{0} \neq \vec{x} \in C_j} \frac{\vec{x}^* A \vec{x}}{\vec{x}^* \vec{x}}$$

for $j = 1, 2, \ldots, n$.

∎

It is often useful to extract spectral information for a sum of matrices by using the individual eigenvalues of the summands. While strict equalities cannot be deduced, bounds do exist. These are contained in the following two theorems, where A has eigenvalues $\alpha_1 \leq \alpha_2 \leq \cdots \leq \alpha_n$, B has eigenvalues $0 < \beta_1 \leq \beta_2 \leq \cdots \leq \beta_n$, and C has eigenvalues $\gamma_1 \leq \gamma_2 \leq \cdots \leq \gamma_n$.

Theorem A.18 (Weyl's Monotonicity Theorem) *Let A and B be Hermitian and $C = A + B$. Then for any i, j satisfying $1 \leq i + j - 1 \leq n$*

$$\alpha_i + \beta_j \leq \gamma_{i+j-1}.$$

In particular, for $j = 1, 2, \ldots, n$,

$$\alpha_1 + \beta_j \leq \gamma_j \leq \alpha_n + \beta_j$$

and

$$\alpha_j + \beta_1 \leq \gamma_j \leq \alpha_j + \beta_n.$$

∎

The most natural development of eigenvalue bounds for *finite Toeplitz forms* (Toeplitz matrices) is via the use of Hermitian matrices and comes from [4, pages 63–65]. If $f \in L^2(-\pi, \pi)$ and is 2π-periodic, then it has the Fourier development

$$f(t) = \sum_{\ell=-\infty}^{\infty} a_\ell e^{i\ell t} \qquad (A.16)$$

where

$$a_\ell = \frac{1}{2\pi} \int_{-\pi}^{\pi} f(t) e^{-i\ell t} dt. \qquad (A.17)$$

To each such f associate the quadratic form (finite $n \times n$ Toeplitz form)

$$\begin{aligned} T_n(f) &\equiv \vec{x}^* T_n \vec{x} \\ &= \sum_{k,j=1,\ldots,n} a_{k-j} x_k x_j \\ &= \vec{x}^* \begin{bmatrix} a_0 & a_1 & a_2 & \cdots & \cdots & a_{n-1} \\ a_{-1} & a_0 & a_1 & \cdots & \cdots & a_{n-2} \\ a_{-2} & a_{-1} & a_0 & \cdots & \cdots & a_{n-3} \\ \vdots & \ddots & \ddots & \ddots & \ddots & \vdots \\ \vdots & \ddots & \ddots & \ddots & \ddots & \vdots \\ a_{-n+1} & \cdots & \cdots & \cdots & \cdots & a_0 \end{bmatrix} \vec{x}. \end{aligned} \qquad (A.18)$$

Note that if f is real-valued, the matrix T_n in (A.18) is Hermitian ($a_{k-j} = \overline{a_{j-k}}$). Also

$$T_n(f) = \frac{1}{2\pi} \int_{-\pi}^{\pi} |\sum_{p=0}^{n} x_p e^{ipt}|^2 f(t) dt,$$

so that if $f(t) \equiv 1$ in (A.16) the matrix T_n in (A.18) is the identity matrix; i.e., $T_n(f) = \sum_{p=0}^{n} |x_p|^2$. This advertises the following theorem.

Theorem A.19 *Let f in (A.16) be a real-valued continuous function on $(-\pi, \pi)$ (which implies that T_n is Hermitian). Then f is nonnegative if and only if the Toeplitz forms $T_n(f)$ are nonnegative definite for all n. Moreover, let $\ell \leq f(x) \leq L$ and denote by $\lambda_j^{(n)}$ ($j = 1, 2, \ldots, n$) the (real) eigenvalues of T_n. Then $\ell \leq \lambda_j^{(n)} \leq L$; $j = 1, 2, \ldots, n$; and $\lim_{n \to \infty} \lambda_1^{(n)} = \ell$ and $\lim_{n \to \infty} \lambda_n^{(n)} = L$.*

■

In fact, if $\epsilon > 0$ then the number of eigenvalues for which $\ell \leq \lambda_j^{(n)} < \ell + \epsilon$ or $L - \epsilon < \lambda_j^{(n)} \leq L$ holds tends to infinity with n. In other words, the eigenvalues cluster at each end. Further, if $f(t) \leq g(t)$ on $(-\pi, \pi)$ the eigenvalues of the associated $n \times n$ Toeplitz matrices satisfy $\lambda_j^{(n)} \leq \mu_j^{(n)}$, where $\mu_j^{(n)}$ is the j-th eigenvalue of $T_n(g)$.

A.3 Block Matrices

The numerical solution of partial differential equations gives rise to matrices having a block structure. The properties of such matrices will be described here.

Definition A.20 A *block (partitioned) matrix* A can be represented as
$$A = \begin{bmatrix} A_{11} & A_{12} & \cdots & A_{1\ell} \\ A_{21} & A_{22} & \cdots & A_{2\ell} \\ \vdots & & & \vdots \\ A_{k1} & A_{k2} & \cdots & A_{k\ell} \end{bmatrix}$$
where each submatrix A_{ij} is of dimension $m_i \times n_j$. A *symmetrically partitioned* square matrix of order n is one with $k = \ell$, and each submatrix A_{ij} is of dimension $n_i \times n_j$. Thus A_{ii} is square for each i.

Block operations are analogous to element operations. Hence block matrices permit addition, scalar multiplication, and matrix multiplication via submatrix operations. The *adjoint* of a block matrix A is given by
$$A^* = \begin{bmatrix} A_{11}^* & A_{21}^* & \cdots & A_{k1}^* \\ A_{12}^* & A_{22}^* & \cdots & A_{k2}^* \\ \vdots & & & \vdots \\ A_{1\ell}^* & A_{2\ell}^* & \cdots & A_{k\ell}^* \end{bmatrix}.$$

Very special block matrices can be built with the following operation.

Definition A.21 Let A be an $m \times n$ matrix and B be a $p \times q$ matrix. The *Kronecker* or *tensor product* of A and B is the $mp \times nq$ matrix
$$A \otimes B \equiv \begin{bmatrix} a_{11}B & a_{12}B & \cdots & a_{1n}B \\ a_{21}B & a_{22}B & \cdots & a_{2n}B \\ \vdots & & & \vdots \\ a_{m1}B & a_{m2}B & \cdots & a_{mn}B \end{bmatrix}. \tag{A.19}$$

Not surprisingly the Kronecker product inherits many properties from A and B [2]. Several relevant properties are summarized in the following theorem.

Theorem A.22 *Let A, B, C, and D be of the necessary dimensions so that the following are all defined. Then*

(i) $(\alpha A) \otimes B = A \otimes (\alpha B) = \alpha(A \otimes B)$ *for α a scalar;*

(ii) $(A + B) \otimes C = A \otimes C + B \otimes C$;

(iii) $A \otimes (B \otimes C) = (A \otimes B) \otimes C$;

(iv) $(A \otimes B)(C \otimes D) = AC \otimes BD$;

(v) $(A \otimes B)^* = A^* \otimes B^*$;

(vi) $\mathrm{rank}(A \otimes B) = [\mathrm{rank}(A)][\mathrm{rank}(B)]$;

(vii) $(A \otimes B)^{-1} = A^{-1} \otimes B^{-1}$;

(viii) *There exists a permutation matrix P such that*

$$B \otimes A = P^T(A \otimes B)P.$$

∎

Two very powerful results relate the eigenstructure and other features of A and B to that of their Kronecker product. These results are given in the following theorems.

Theorem A.23 *Let the eigenpairs of A be $\{\alpha_i, \vec{x}_i\}$ and those of B be $\{\beta_j, \vec{y}_j\}$. Then the eigenpairs of $A \otimes B$ are $\{\alpha_i \beta_j, \vec{x}_i \otimes \vec{y}_j\}$.*

∎

Theorem A.24 *If A and B are both*

(i) *normal,*

(ii) *Hermitian,*

(iii) *positive definite,*

(*iv*) *unitary*,

then so is $A \otimes B$.

∎

Two important matrices arise in the numerical solution of partial differential equations. They are described in the following definitions.

Definition A.25 Let A and B be $m \times n$ matrices. The *Hadamard product* of A and B is the $m \times n$ matrix $A \circ B = (a_{ij}b_{ij})$. This is the element-by-element product of two matrices. Hence

$$A \circ B = \begin{bmatrix} a_{11}b_{11} & a_{12}b_{12} & \cdots & a_{1n}b_{1n} \\ a_{21}b_{21} & a_{22}b_{22} & \cdots & a_{2n}b_{2n} \\ \vdots & & & \vdots \\ a_{m1}b_{m1} & a_{m2}b_{m2} & \cdots & a_{mn}b_{mn} \end{bmatrix}.$$

Definition A.26 Let A be $m \times m$ and B be $n \times n$. The *Kronecker sum* of A and B is the $mn \times mn$ matrix

$$\mathcal{A} = I_m \otimes B + A \otimes I_n. \qquad (A.20)$$

The Kronecker sum of A and B can be visualized as

$$\begin{aligned} \mathcal{A} &= I_m \otimes B + A \otimes I_n \\ &= \begin{bmatrix} a_{11}I_n + B & a_{12}I_n & \cdots & a_{1m}I_n \\ a_{21}I_n & a_{22}I_n + B & \cdots & a_{2m}I_n \\ \vdots & \ddots & \ddots & \vdots \\ \vdots & \ddots & \ddots & \vdots \\ a_{m1}I_n & a_{m2}I_n & \cdots & a_{mm}I_n + B \end{bmatrix} \end{aligned}$$

where each submatrix C_{ij} is given by

$$C_{ij} = a_{ij}I_n + \delta_{ij}^{(0)} B. \qquad (A.21)$$

Here $\delta_{ij}^{(0)}$ is the Kronecker delta.

Theorem A.27 *Let the eigenpairs of A be $\{\alpha_i, \vec{x}_i\}$ and those of B be $\{\beta_j, \vec{y}_j\}$. Then the eigenpairs of the Kronecker sum $I_m \otimes B + A \otimes I_n$ are $\{(\alpha_i + \beta_j), \vec{x}_i \otimes \vec{y}_j\}$.*

∎

Theorem A.28 *If A and B are both*

(i) *normal,*

(ii) *Hermitian,*

(iii) *positive definite,*

then so is the Kronecker sum of A and B.

∎

Lastly, two operations are defined that provide an organized way to represent discrete systems arising from multidimensional problems. The definition of the *concatenation* operation used here differs slightly from that in [2, page 193].

Definition A.29 If b is a one-dimensional array $b = (b_i)$, $1 \leq i \leq m$, then the *concatenation* of b is the vector

$$\mathrm{co}(b) \equiv \begin{bmatrix} b_1 \\ b_2 \\ \vdots \\ b_m \end{bmatrix}.$$

Similarly, if $B = (b_{ij})$, $1 \leq i \leq m$, $1 \leq j \leq n$, is a two-dimensional array, then the *concatenation* of B is the $mn \times 1$ vector

$$\mathrm{co}(B) = \begin{bmatrix} \mathrm{co}(b_{i1}) \\ \mathrm{co}(b_{i2}) \\ \vdots \\ \mathrm{co}(b_{in}) \end{bmatrix}.$$

In general, if $B = (b_{i_1 i_2 \ldots i_n})$, $1 \leq i_j \leq m_j$, $1 \leq j \leq n$, is an n-dimensional array, then the *concatenation* of B is recursively defined by

$$\operatorname{co}(B) = \operatorname{co}((b_{i_1 i_2 \ldots i_n}))$$

$$\equiv \begin{bmatrix} \operatorname{co}((b_{i_1 i_2 \ldots i_{n-1} 1})) \\ \operatorname{co}((b_{i_1 i_2 \ldots i_{n-1} 2})) \\ \vdots \\ \operatorname{co}((b_{i_1 i_2 \ldots i_{n-1} m_n})) \end{bmatrix}.$$

Thus $\operatorname{co}(B)$ is a vector of dimension $\prod_{j=1}^{n} m_j \times 1$.

The *matrix* operation, "mat," allows the representation of any n-dimensional ($n \geq 2$) array as a matrix. The following definition provides the details.

Definition A.30 If $B = (b_{ij})$, $1 \leq i \leq m$, $1 \leq j \leq n$, is a two-dimensional array, then the *matrix* of B is the $m \times n$ matrix

$$\operatorname{mat}(B) = [\operatorname{co}(b_{i1}), \operatorname{co}(b_{i2}), \ldots, \operatorname{co}(b_{in})].$$

Similarly, if $B = (b_{ij\ell})$, $1 \leq i \leq m$, $1 \leq j \leq n$, $1 \leq \ell \leq p$, is a three-dimensional array, then the *matrix* of B is the $mn \times p$ matrix

$$\operatorname{mat}(B) = \operatorname{mat}((b_{ij\ell}))$$
$$\equiv [\operatorname{co}(b_{ij1}), \operatorname{co}(b_{ij2}), \ldots, \operatorname{co}(b_{ijp})].$$

In general, if $B = (b_{i_1 i_2 \ldots i_n})$, $1 \leq i_j \leq m_j$, $1 \leq j \leq n$, is an n-dimensional array, then the *matrix* of B is recursively defined by

$$\operatorname{mat}(B) = \operatorname{mat}((b_{i_1 i_2 \ldots i_n}))$$
$$\equiv [\operatorname{co}((b_{i_1 \ldots i_{n-1} 1})), \operatorname{co}((b_{i_1 \ldots i_{n-1} 2})), \ldots, \operatorname{co}((b_{i_1 \ldots i_{n-1} m_n}))].$$

Thus the matrix operation unravels the first $n-1$ indices of the array leaving a matrix with $\prod_{j=1}^{n-1} m_j$ rows and m_n columns.

A useful property of concatenation is given in the following theorem.

Theorem A.31 *If A and B are arrays of identical dimension and α and β are scalars, then*

$$\mathrm{co}(\alpha A + \beta B) = \alpha \mathrm{co}(A) + \beta \mathrm{co}(B).$$

∎

To unravel the discrete systems that arise (and assist in their solution), the following theorem is used extensively.

Theorem A.32 *Let A be $m \times m$, X be $m \times n$, and B be $n \times n$. Then*

$$\mathrm{co}(AXB) = (B^T \otimes A)\mathrm{co}(X). \tag{A.22}$$

∎

This leads to the following general equivalence.

Theorem A.33 *The linear system for the unknown matrix X is given as*
$$A_1 X B_1 + A_2 X B_2 + \cdots + A_k X B_k = C \tag{A.23}$$
where A_i are $m \times m$; X, C are $m \times n$; and B_i are $n \times n$. It is equivalent to
$$G \mathrm{co}(X) = \mathrm{co}(C) \tag{A.24}$$
where
$$G \equiv B_1^T \otimes A_1 + B_2^T \otimes A_2 + \cdots + B_k^T \otimes A_k. \tag{A.25}$$

∎

The equivalence given in Theorem A.33 provides two ways to approach the solution of a problem posed as (A.23). One can tackle this problem directly or use its equivalent formulation (A.24). The reduction or simplification of either formulation often has an analogue for the other. For example, the diagonalization of G in (A.25) can be accomplished via the simultaneous diagonalization of pairs of the components of G. As an illustration, consider (all matrices $n \times n$)

$$A_1 X B_1 + A_2 X B_2 = C. \tag{A.26}$$

LINEAR ALGEBRA

Assume that A_1 and B_2 are symmetric with eigenvalues $\{\alpha_i\}_{i=1}^n$ and $\{\beta_j\}_{j=1}^n$, respectively, and that A_2 and B_1 are positive definite. Then (A.26) is equivalent to

$$G\text{co}(X) = \text{co}(C) \qquad (A.27)$$

where

$$G = B_1 \otimes A_1 + B_2 \otimes A_2.$$

Theorem A.13 guarantees matrices S_1 and S_2 that simultaneously diagonalize the pairs (A_1, A_2) and (B_2, B_1), respectively. So

$$S_1^T A_1 S_1 = D(\alpha), \quad S_1^T A_2 S_1 = I$$

and

$$S_2^T B_2 S_2 = D(\beta), \quad S_2^T B_1 S_2 = I.$$

Then (A.26) becomes

$$D(\alpha) Y + Y D(\beta) = F \qquad (A.28)$$

where

$$Y = S_1^{-1} X (S_2^T)^{-1}$$

and

$$F = S_1^T C S_2.$$

Equivalently, (A.27) may be written

$$[I \otimes D(\alpha) + D(\beta) \otimes I]\text{co}(Y) = \text{co}(F). \qquad (A.29)$$

If each of the pairs (A_1, A_2) and (B_2, B_1) are not simultaneously diagonalizable, a generalization of Schur's Theorem A.6 still provides a reduction. This is contained in the next theorem.

Theorem A.34 (Generalized Schur Theorem) *Let A and B be $n \times n$ matrices. Then there exist unitary matrices Q and R such that*

$$Q^* A R = T_1, \quad Q^* B R = T_2$$

where T_1 and T_2 are upper triangular.

∎

Thus the problem in (A.26) for arbitrary matrices A_1, A_2, B_1, and B_2 could be reduced as follows. By the previous theorem there exist unitary matrices Q_1, R_1, Q_2, and R_2 so that

$$Q_1^* A_1 R_1 = T_1, \quad Q_1^* A_2 R_1 = T_2$$

and

$$Q_2^* B_1^* R_2 = T_3, \quad Q_2^* B_2^* R_2 = T_4.$$

Then (A.26) reduces to

$$T_1 Y T_3^* + T_2 Y T_4^* = F \qquad (A.30)$$

where

$$Y = R_1^* X Q_2$$

and

$$F = Q_1^* C R_2.$$

The special system (*Sylvester's equation*)

$$AX + XB = C, \qquad (A.31)$$

which often arises in numerical partial differential equations, is equivalent to the system

$$(I_n \otimes A + B^T \otimes I_m)\text{co}(X) = \text{co}(C) \qquad (A.32)$$

which involves the Kronecker sum. If A and B are diagonalizable via the matrices P and Q, respectively, then Sylvester's equation may be written

$$D(\alpha)Y + Y D(\beta) = F \qquad (A.33)$$

where

$$Y = P^{-1} X Q$$

and

$$F = P^{-1} C Q.$$

Hence the solution of the transformed Sylvester's equation (A.33) is given by

$$y_{i,j} = \frac{1}{\alpha_i + \beta_j}$$

and X is recovered from Y via $X = PYQ^{-1}$.

References

[1] R. Bellman, *Introduction to Matrix Analysis*, McGraw-Hill, Inc., New York, 1960.

[2] P. J. Davis, *Circulant Matrices*, John Wiley & Sons, Inc., New York, 1979.

[3] G. H. Golub and C. F. Van Loan, *Matrix Computations*, Second Ed., Johns Hopkins University Press, Baltimore, 1989.

[4] V. Grenander and G. Szego, *Toeplitz Forms and Their Applications*, Second Ed., Chelsea Publishing Co., New York, 1984.

[5] P. Lancaster and M. Tismenetsky, *The Theory of Matrices*, Second Ed., Academic Press, Inc., Orlando, 1985.

[6] B. N. Parlett, *The Symmetric Eigenvalue Problem*, Prentice-Hall, Inc., Englewood Cliffs, 1980.

[7] J. H. Wilkinson, *The Algebraic Eigenvalue Problem*, Oxford University Press, Oxford, 1965.

Bibliography

L. V. Ahlfors, *Complex Analysis*, Third Ed., McGraw-Hill, Inc., New York, 1979.

H. T. Banks and K. Kunisch, *Estimation Techniques for Distributed Parameter Systems*, Birkhäuser, Boston, 1989.

R. Bellman, *Introduction to Matrix Analysis*, McGraw-Hill, Inc., New York, 1960.

C. M. Bender and S. A. Orszag, *Advanced Mathematical Methods for Scientists and Engineers*, McGraw-Hill, Inc., New York, 1978.

B. Bialecki, "Sinc-Type Approximations in H^1-Norm with Application to Boundary Value Problems," *J. Comput. Appl. Math.*, 25 (1989), pages 289–303.

B. Bialecki, "Sinc-Collocation Methods for Two-Point Boundary Value Problems," *IMA J. Numer. Anal.*, 11 (1991), pages 357–375.

K. L. Bowers and J. Lund, "Numerical Solution of Singular Poisson Problems via the Sinc-Galerkin Method," *SIAM J. Numer. Anal.*, 24 (1987), pages 36–51.

J. D. Cole, "On a Quasi-Linear Parabolic Equation Occurring in Aerodynamics," *Quarterly Appl. Math.*, 9 (1951), pages 225–236.

P. J. Davis, *Circulant Matrices*, John Wiley & Sons, Inc., New York, 1979.

P. J. Davis and P. Rabinowitz, *Methods of Numerical Integration*, Second Ed., Academic Press, Inc., San Diego, 1984.

R. R. Doyle, "Extensions to the Development of the Sinc-Galerkin Method for Parabolic Problems," Ph.D. Thesis, Montana State University, Bozeman, MT, 1990.

N. Eggert, M. Jarratt, and J. Lund, "Sinc Function Computation of the Eigenvalues of Sturm–Liouville Problems," *J. Comput. Phys.*, 69 (1987), pages 209–229.

N. Eggert and J. Lund, "The Trapezoidal Rule for Analytic Functions of Rapid Decrease," *J. Comput. Appl. Math.*, 27 (1989), pages 389–406.

D. S. Gilliam, J. R. Lund, and C. F. Martin, "A Discrete Sampling Inversion Scheme for the Heat Equation," *Numer. Math.*, 54 (1989), pages 493–506.

G. H. Golub, S. Nash, and C. Van Loan, "A Hessenberg–Schur Method for the Problem $AX + XB = C$," *IEEE Trans. Automat. Control*, 24 (1979), pages 909–913.

G. H. Golub and C. F. Van Loan, *Matrix Computations*, Second Ed., Johns Hopkins University Press, Baltimore, 1989.

I. S. Gradshteyn and I. M. Ryzhik, *Table of Integrals, Series and Products*, Academic Press, Inc., New York, 1980.

V. Grenander and G. Szego, *Toeplitz Forms and Their Applications*, Second Ed., Chelsea Publishing Co., New York, 1984.

S.-Å. Gustafson and F. Stenger, "Convergence Acceleration Applied to Sinc Approximation with Application to Approximation of $|x|^\alpha$," in *Computation and Control II*, Proc. Bozeman Conf. 1990, Progress in Systems and Control Theory, Vol. 11, Birkhäuser, Boston, 1991, pages 161–171.

E. Hille, *Analytic Function Theory*, Vol. II, Second Ed., Chelsea Publishing Co., New York, 1977.

M. Jarratt, "Approximation of Eigenvalues of Sturm–Liouville Differential Equations by the Sinc-Collocation Method," Ph.D. Thesis, Montana State University, Bozeman, MT, 1987.

M. Jarratt, "Eigenvalue Approximations on the Entire Real Line," in *Computation and Control*, Proc. Bozeman Conf. 1988, Progress in Systems and Control Theory, Vol. 1, Birkhäuser, Boston, 1989, pages 133–144.

M. Jarratt, "Eigenvalue Approximations for Numerical Observability Problems," in *Computation and Control II*, Proc. Bozeman Conf. 1990, Progress in Systems and Control Theory, Vol. 11, Birkhäuser, Boston, 1991, pages 173–185.

M. Jarratt, J. Lund, and K. L. Bowers, "Galerkin Schemes and the Sinc-Galerkin Method for Singular Sturm–Liouville Problems," *J. Comput. Phys.*, 89 (1990), pages 41–62.

P. Lancaster and M. Tismenetsky, *The Theory of Matrices*, Second Ed., Academic Press, Inc., Orlando, 1985.

D. L. Lewis, "A Fully Galerkin Method for Parabolic Problems," Ph.D. Thesis, Montana State University, Bozeman, MT, 1989.

D. L. Lewis, J. Lund, and K. L. Bowers, "The Space-Time Sinc-Galerkin Method for Parabolic Problems," *Internat. J. Numer. Methods Engrg.*, 24 (1987), pages 1629–1644.

J. Lund, "Sinc Function Quadrature Rules for the Fourier Integral," *Math. Comp.*, 41 (1983), pages 103–113.

J. Lund, "Symmetrization of the Sinc-Galerkin Method for Boundary Value Problems," *Math. Comp.*, 47 (1986), pages 571–588.

J. Lund, "Accuracy and Conditioning in the Inversion of the Heat Equation," in *Computation and Control*, Proc. Bozeman Conf. 1988, Progress in Systems and Control Theory, Vol. 1, Birkhäuser, Boston, 1989, pages 179–196.

J. Lund, "Sinc Approximation Method for Coefficient Identification in Parabolic Systems," in *Robust Control of Linear Systems and Nonlinear Control, Vol. II*, Proc. MTNS 1989, Progress in Systems and Control Theory, Vol. 4, Birkhäuser, Boston, 1990, pages 507–514.

J. Lund, K. L. Bowers, and T. S. Carlson, "Fully Sinc-Galerkin Computation for Boundary Feedback Stabilization," *J. Math. Systems, Estimation, and Control*, 1 (1991), pages 165–182.

J. Lund, K. L. Bowers, and K. M. McArthur, "Symmetrization of the Sinc-Galerkin Method with Block Techniques for Elliptic Equations," *IMA J. Numer. Anal.*, 9 (1989), pages 29–46.

J. Lund and B. V. Riley, "A Sinc-Collocation Method for the Computation of the Eigenvalues of the Radial Schrödinger Equation," *IMA J. Numer. Anal.*, 4 (1984), pages 83–98.

J. Lund and C. R. Vogel, "A Fully-Galerkin Method for the Numerical Solution of an Inverse Problem in a Parabolic Partial Differential Equation," *Inverse Problems*, 6 (1990), pages 205–217.

L. Lundin, "A Cardinal Function Method of Solution of the Equation $\Delta u = u - u^3$," *Math. Comp.*, 35 (1980), pages 747–756.

L. Lundin and F. Stenger, "Cardinal Type Approximations of a Function and Its Derivatives," *SIAM J. Math. Anal.*, 10 (1979), pages 139–160.

J. E. Marsden and M. J. Hoffman, *Basic Complex Analysis*, Second Ed., W. H. Freeman and Co., New York, 1987.

K. M. McArthur, "Sinc-Galerkin Solution of Second-Order Hyperbolic Problems in Multiple Space Dimensions," Ph.D. Thesis, Montana State University, Bozeman, MT, 1987.

K. M. McArthur, "A Collocative Variation of the Sinc-Galerkin Method for Second Order Boundary Value Problems," in *Computation and Control*, Proc. Bozeman Conf. 1988, Progress in Systems and Control Theory, Vol. 1, Birkhäuser, Boston, 1989, pages 253–261.

K. M. McArthur, K. L. Bowers, and J. Lund, "Numerical Implementation of the Sinc-Galerkin Method for Second-Order Hyperbolic Equations," *Numer. Methods for Partial Differential Equations*, 3 (1987), pages 169–185.

K. M. McArthur, K. L. Bowers, and J. Lund, "The Sinc Method in Multiple Space Dimensions: Model Problems," *Numer. Math.*, 56 (1990), pages 789–816.

K. M. McArthur, R. C. Smith, J. Lund, and K. L. Bowers, "The Sinc-Galerkin Method for Parameter Dependent Self-Adjoint Problems," accepted by *Appl. Math. Comput.*

J. McNamee, F. Stenger, and E. L. Whitney, "Whittaker's Cardinal Function in Retrospect," *Math. Comp.*, 25 (1971), pages 141–154.

S. G. Mikhlin, *Variational Methods in Mathematical Physics*, Pergamon Press Ltd., Oxford, 1964.

I. P. Natanson, *Constructive Function Theory, Vol. I: Uniform Approximation*, Frederick Ungar Publishing Co., Inc., New York, 1964.

F. W. J. Olver, *Asymptotics and Special Functions*, Academic Press, Inc., New York, 1974.

B. N. Parlett, *The Symmetric Eigenvalue Problem*, Prentice-Hall, Inc., Englewood Cliffs, 1980.

P. M. Prenter, *Splines and Variational Methods*, John Wiley & Sons, Inc., New York, 1975.

B. V. Riley, "Galerkin Schemes for Elliptic Boundary Value Problems," Ph.D. Thesis, Montana State University, Bozeman, MT, 1982.

W. Rudin, *Real and Complex Analysis*, Third Ed., McGraw-Hill, Inc., New York, 1987.

S. Schaffer and F. Stenger, "Multigrid-Sinc Methods," *Appl. Math. Comput.*, 19 (1986), pages 311–319.

J. L. Schwing, "Numerical Solutions of Problems in Potential Theory," Ph.D. Thesis, University of Utah, Salt Lake City, UT, 1976.

R. C. Smith, "Numerical Solution of Fourth-Order Time-Dependent Problems with Applications to Parameter Identification," Ph.D. Thesis, Montana State University, Bozeman, MT, 1990.

R. C. Smith, G. A. Bogar, K. L. Bowers, and J. Lund, "The Sinc-Galerkin Method for Fourth-Order Differential Equations," *SIAM J. Numer. Anal.*, 28 (1991), pages 760–788.

R. C. Smith and K. L. Bowers, "A Fully Galerkin Method for the Recovery of Stiffness and Damping Parameters in Euler–Bernoulli Beam Models," in *Computation and Control II*, Proc. Bozeman Conf. 1990, Progress in Systems and Control Theory, Vol. 11, Birkhäuser, Boston, 1991, pages 289–306.

R. C. Smith, K. L. Bowers, and J. Lund, "Efficient Numerical Solution of Fourth-Order Problems in the Modeling of Flexible Structures," in *Computation and Control*, Proc. Bozeman Conf. 1988, Progress in Systems and Control Theory, Vol. 1, Birkhäuser, Boston, 1989, pages 283–297.

R. C. Smith, K. L. Bowers, and J. Lund, "A Fully Sinc-Galerkin Method for Euler–Bernoulli Beam Models," *Numer. Methods Partial Differential Equations*, 8 (1992), pages 171–202.

R. C. Smith, K. L. Bowers, and C. R. Vogel, "Numerical Recovery of Material Parameters in Euler–Bernoulli Beam Models," accepted by *J. Math. Systems, Estimation, and Control*

F. Stenger, "Integration Formulae Based on the Trapezoidal Formula," *J. Inst. Maths. Appl.*, 12 (1973), pages 103–114.

F. Stenger, "Approximations via Whittaker's Cardinal Function," *J. Approx. Theory*, 17 (1976), pages 222–240.

F. Stenger, "A Sinc-Galerkin Method of Solution of Boundary Value Problems," *Math. Comp.*, 33 (1979), pages 85–109.

F. Stenger, "Numerical Methods Based on Whittaker Cardinal, or Sinc Functions," *SIAM Rev.*, 23 (1981), pages 165–224.

M. Stromberg, "Solution of Shock Problems by Methods Using Sinc Functions," Ph.D. Thesis, University of Utah, Salt Lake City, UT, 1988.

M. Stromberg, "Sinc Approximate Solution of Quasilinear Equations of Conservation Law Type," in *Computation and Control*, Proc. Bozeman Conf. 1988, Progress in Systems and Control Theory, Vol. 1, Birkhäuser, Boston, 1989, pages 317–331.

E. C. Titchmarsh, *Theory of Fourier Integrals*, Clarendon Press, Oxford, 1954.

E. T. Whittaker, "On the Functions Which Are Represented by the Expansions of the Interpolation-Theory," *Proc. Roy. Soc. Edinburgh*, 35 (1915), pages 181–194.

J. M. Whittaker, *Interpolatory Function Theory*, Cambridge Tracts in Mathematics and Mathematical Physics, No. 33, Cambridge University Press, London, 1935.

J. H. Wilkinson, *The Algebraic Eigenvalue Problem*, Oxford University Press, Oxford, 1965.

Text Directory

Analytic Functions, 1–17
    Def. 1.11 Sinc Function, 5
    Sinc $\in L^2(\mathbb{R})$, Ex. 1.24, 14
    The Fourier Transform of
        the Sinc Function, 15
    Sinc Orthogonality, 16
    Paley–Weiner Theorem, 17

    Exercises, 17–18

Exact Interpolation $\mathbb{R}$, 22–24
    The Class $B(h)$, 22
    Reproducing Kernel, 23
    Sinc Series, 24

Exact Quadrature $\mathbb{R}$, 24–25
    Trapezoidal Rule, 24
    Orthonormal Set, 25

Intuitive Examples, 27–33
    Observability,
        Ex. 2.9, 27–28
    Interpolation $(x^2+1)^{-1}$,
        Ex. 2.10, 29–31
    The Trapezoidal Rule on
        $\int e^{-|t|}dt$ vs. $\int \text{sech}(t)dt$,
        Ex. 2.11, 31–33

Interpolation, 33–46
    D_S (Strip), 34
    Interpolation of f, 35–40
    Mesh Restriction, 40
    Interpolation of $f^{(n)}$, 41–45
    Collocation,
        Ex. 2.17, 45–46

Quadrature, 47–51
    Trapezoidal Rule, 48–50
    The Sinc-Galerkin Method,
        Ex. 2.22, 50–51

    Exercises, 52–54

Def. 3.1 The Class $B(D)$, 58–59

Interpolation on $\Gamma \subset D$, 59–68
    $D = D_E$ (Eyeball) $(0,1)$, 63
    Interpolation $x^{1/3}(1-x)^{1/2}$,
        Ex. 3.6, 66–68

Quadrature on Γ, 68–90
    $D = D_W$ (Wedge) $(0,\infty)$,
        Nonoscillatory, 72
    $\int_0^\infty \frac{du}{u^2+1}$,
        Bounds on Wedge, 73–75

$\int_0^\infty [\frac{1}{u+1} - e^{-u}]\frac{du}{u}$,
  Ex. 3.10, 75–76
Economization, 77–78
$\int_0^\infty u^{\nu-1} e^{-u} du$,
  Ex. 3.11, 78–79
$D = D_B$ (Bullet) $(0, \infty)$,
  Oscillatory, 79–83
$\int_0^\infty \exp(-u)\sin(\lambda u) du$,
  Ex. 3.13, 83–87
$D = D_H$ (Hyperbolic
  Region) $(-\infty, \infty)$, 88
$\int_{\mathbf{R}} \frac{du}{u^4 + a^4}$,
  Ex. 3.15, 89–90

Table 3.7: Weights and Nodes
for Quadrature, 91

Differentiation on Γ, 92–96
  Interpolation of
  $u^{3/2}(1-u)\ell n(u)$,
  Ex. 3.18, 95–96

Exercises, 97–98

Galerkin Schemes, 102–103

Inner Product, 104–113

System Assembly, 113–115
  $I^{(1)}$ and $I^{(2)}$, 115

$Lu = -u'' + p(x)u' + q(x)u = f$,
  Discrete System, 115–117

Table 4.1: System Entries, 118

Examples, 119–127
  $Lu = -u'' - \frac{u'}{6x} + \frac{u}{x^2}$,
  Ex. 4.5 $(0, 1)$, 119–120

$Lu = -u'' + \kappa u'$,
  Ex. 4.6 $(0, 1)$, 120–122
$Lu = -u'' + \frac{u}{x^2}$,
  Ex. 4.8 $(0, 1)$, 124–125
$Lu = -u'' + u$,
  Ex. 4.9 $(0, \infty)$, 125–126
$Lu = -u'' + \frac{xu'}{x^2+1} + \frac{u}{x^2+1}$,
  Ex. 4.10 $(0, \infty)$, 126–127

$L_s u = -u'' + q(x)u = f$
  Discrete System, 129–131

Table 4.7: System Entries, 131

Examples, 133–136
  $L_s u = -u'' + \frac{u}{x^2}$,
  Ex. 4.12 $(0, 1)$, 133–134
  $L_s u = -u'' + \frac{3u}{4x^2}$,
  Ex. 4.13 $(0, \infty)$, 134–136

Initial Value Problem, 136–140
  $u' + \frac{3u}{2x} = f$,
  Ex. 4.14 $(0, \infty)$, 137–138
  $u' - u = f$,
  Ex. 4.15 $(0, \infty)$, 139–140

Radiation Boundary
  Conditions, 140–150
  Boundary Basis
    Functions, 142
  Discrete System, 146
  $Lu = -u'' + \frac{u'}{x} + \frac{u}{x^2}$,
    Ex. 4.16 $(0, 1)$, 147–148
  $Lu = -u'' + \kappa u'$,
    Ex. 4.17 $(0, 1)$, 148–150
System Invertibility and
  Implementation, 151–154

Exercises, 155–160

Sinc-Collocation, 164–173
$Lu = -u'' - \frac{u'}{6x} + \frac{u}{x^2}$,
Ex. 5.1 $(0,1)$, 172–173

Sturm–Liouville, 174–188
Harmonic Oscillator,
Ex. 5.3, 182–184
Hermite's Equation,
Ex. 5.4, 184–188
Matrix Splitting, 185–186

Self-Adjoint Form, 188–194
Ex. 5.5, 192–194

Poisson Problem, 194–218
2-Dimensional, 194–209
Kronecker Form, 199
1-Dimensional Point
of View, 200–202
Concatenated Form, 203
Diagonalization Solution
Method, 204–205
Schur's Method, 207
Ex. 5.6, 208–209
3-Dimensional, 210–216
Kronecker Form, 211
Concatenated Form, 213
Ex. 5.7, 215–216

Unique Solvability, 216–218

Exercises, 218–222

Heat Equation, 228–254
1-Dimensional, 228–237
Kronecker Form, 231
Concatenated Form, 233
Ex. 6.2, 235–237
2-Dimensional, 237–241
Kronecker Form, 238
Concatenated Form, 239
Ex. 6.3, 240–241
Solvability, 241–242
General Boundary
Conditions, 242–254
Ex. 6.4, 249–250
Ex. 6.5, 250–254

Transport Problem, 254–263
Advection-Diffusion, 254
Ex. 6.6, 256–257
Burgers' Equation, 257
Ex. 6.7, 260–263

Exercises, 263–266

Index

$B(D)$, 58
$B(h)$, 22
Basis Functions
   $(0,1)$ Fig. 3.4, 66
   $(0,\infty)$ Fig. 3.6, 74
   $(0,\infty)$ Fig. 3.8, 83
   $(0,1) \times (0,1)$ Fig. 5.1, 196
   $(0,1) \times (0,\infty)$ Fig. 6.1, 229
   $(-\infty,\infty)$ Fig. 2.1, 26
   $(-\infty,\infty)$ Fig. 3.10, 89
Boundary Value Problem
   Fourth-Order, 155–157
   Poisson Problem, 194–218
   Second-Order, 104–135
      Collocation System, 170
      Galerkin System, 116

Derivatives, 41, 93
Domains and Conformal Maps
   $(0,1)$ D_E, 63–65
   $(0,\infty)$ D_W, 71–72
   $(0,\infty)$ D_B, 80–81
   $(-\infty,\infty)$ D_S, 34
   $(-\infty,\infty)$ D_H, 87–88

Eigenvalues of
   $I^{(1)}$, $I^{(2)}$, 151–152

Heat Equation, 228–254

Initial Value Problem, 136–140
Inner Products, 104–113
Interpolation
   Approximate on $\mathbb{R}$, 38
   Approximate on Γ, 59
   Exact on $\mathbb{R}$, 24

Matrices
   $I^{(1)}$, 106, 115
   $I^{(2)}$, 106, 115
   $I^{(3)}$, 155, 157
   $I^{(4)}$, 156, 157
   Concatenation, 284
   Kronecker Sum, 283
   Mat Operator, 285
   Toeplitz, 277, 280

Nonhomogeneous Conditions
   Heat Equation, 242–254
   Scalar Problem, 140–150
   Discrete System, 146

Paley–Wiener Theorem, 17

Quadrature
   Approximate on $\mathbb{R}$, 48–51
   Approximate on Γ, 69–71
   Exact on $\mathbb{R}$, 24

Self-Adjoint Form, 188–194
Sinc Function
    Composed with Maps
        (Figures), 66, 74, 83, 89
    Definition, 5
    Fourier Transform, 15
    On $\mathbb{R}$, 26
    Orthonormal Set, 25
    Reproducing Kernel, 23
Sturm–Liouville, 174–188

Tables
    3.7: Weights and Nodes for
        Quadrature, 91
    4.1: $-u'' + pu' + qu = f$
        (Discrete System), 118
    4.7: $-u'' + qu = f$
        (Symmetric System), 131
Transport Problems, 254–263
    Advection-Diffusion, 254
    Burgers' Equation, 257